The hunger for sodium has been used as a model system in which to study how the brain produces motivated behavior. In this account of the field, Jay Schulkin draws together information across a range of disciplines and topics, ranging from the ecology of salt ingestion to the sodium molecule and the action of various hormones. The phenomenon of sodium hunger was discovered by Curt Richter, the great American psychobiologist, over 50 years ago. Its study has been of interest for some time to naturalists, psychologists, endocrinologists, physiologists and neuroscientists. This book offers a systematic account of the behavior of the sodium hungry animal as well as the endocrine and physiological mechanisms that act to maintain sodium balance and act on the brain to promote the search for, and the ingestion of, salt. Finally, the book provides a description of a neural network that orchestrates the behavior of salt seeking and salt ingestion.

Sodium hunger: the search for a salty taste

Sodium hunger:
the search for a salty taste

JAY SCHULKIN
Department of Anatomy and Institute of Neurological Sciences
University of Pennsylvania

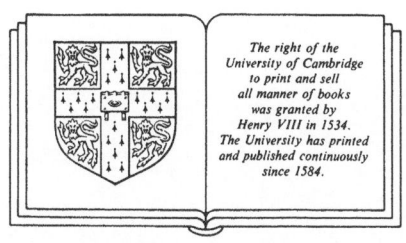

The right of the
University of Cambridge
to print and sell
all manner of books
was granted by
Henry VIII in 1534.
The University has printed
and published continuously
since 1584.

CAMBRIDGE UNIVERSITY PRESS
Cambridge
New York   Port Chester   Melbourne   Sydney

CAMBRIDGE UNIVERSITY PRESS
Cambridge, New York, Melbourne, Madrid, Cape Town, Singapore, São Paulo

Cambridge University Press
The Edinburgh Building, Cambridge CB2 2RU, UK

Published in the United States of America by Cambridge University Press, New York

www.cambridge.org
Information on this title: www.cambridge.org/9780521353687

First published 1991
This digitally printed first paperback version 2005

A catalogue record for this publication is available from the British Library

Library of Congress Cataloguing in Publication data
Schulkin, Jay.
Sodium hunger: the search for a salty taste / Jay Schulkin.
    p.   cm.
Includes bibliographical references and index.
ISBN 0-521-35368-8
1. Salt in the body.   2. Appetite.   I. Title.
QP535. N2S38   1991
612.3′926–dc20   91–11018 CIP

ISBN-13 978-0-521-35368-7 hardback
ISBN-10 0-521-35368-8 hardback

ISBN-13 978-0-521-01842-5 paperback
ISBN-10 0-521-01842-0 paperback

To my science teacher, George Wolf

Contents

Preface   *page* xi

Introduction   1

1   Salt seeking behavior   6

2   Hormonal regulation of salt intake   31

3   Gustatory contribution to salt intake   57

4   Physiological factors in the control of salt intake   87

5   Neural circuits underlying salt intake   110

Conclusion   138

Appendix   140

References   141

Name index   183

Subject index   189

Preface

When I was an undergraduate philosophy student, I heard from other students that the mind–body problem was really an issue of how the brain works. At my school, which was largely an art school, there was a neuroscientist named George Wolf. From him, I learned how the brain works. He had an interest in philosophy, so I could reciprocate and perhaps engage and nurture him in his own philosophical interests.

Within a short period of time, he handed me my first scientific project. It was not about the brain, though it quickly evolved into a project that was, because it was about a simple behavior – salt ingestion. The question was whether one hormone that generates this behavior does so by innate, as opposed to learned, mechanisms.

I was very fortunate: my first project worked. I found that mineralocorticoid-induced sodium hunger is innate. George, despite the fact that he told me what to do and how to do it (while letting me think I was doing it myself) declined to put his name on the paper. He told me that 'if I put my name on the paper, you won't get any credit'. In fact, it was the only paper that I have ever had accepted unconditionally.

Despite my going off to the University of Pennsylvania, George continued to be my scientific advisor, close friend and collaborator. But, at Penn and other places, I discovered and engaged a number of people who have enriched my life enormously. In fact, I have felt that I have had the best colleagues that one could hope for. I list some of them for you now: Gary Beauchamp, Linda Bartoshuk, Kent Berridge, Ted Coons, Derek Denton, Alan Epstein, Steve Fluharty, Bill Flynn, Mark Friedman, Harvey Grill, E.E. Krieckhaus, Maurizio Mass, Bruce McEwen, Rich Miselis, Ralph Norgren, Alan Rosenwasser, Paul Rozin, John Sabini, David Sarokin, Eliot Stellar and Mike Tordoff.

The data presented here reflect the laboratories I have worked in, colleagues I have been close to, and research findings I know the best. For those who, perhaps, may feel slighted I apologize. Authors, and I'm no different, stay with what they are most familiar: the work they have been involved in, or the people and research that they know well. Another factor I would like to call attention to is that this book emphasizes the behavioral side of sodium homeostasis. Thus, for a greater emphasis on physiological and pathological parameters, I refer the reader to Derek Denton's fine book on the subject.

Finally, I thank my friends, family, Bob and April for their encouragement and support. I have been supported by a Research Career Development Award from the National Institute of Mental Health, 00678, and by a Program Project Grant from the National Institute of Mental Health, 43787.

Introduction

At first glance, one might ask what the hunger for sodium has to do with psychology. In fact, most psychology textbooks hardly mention it. By contrast, learning and motivation are typical psychological topics. Hunger and thirst, and the study of stress and the emotions or conceptions of mind, are clearly included under the purview of psychology. What I hope to prove is that the hunger for sodium – or the search for something salty, which is expressed in a variety of different animals – is an ideal behavioral system to study, and is an appropriate topic within psychology.

INTELLECTUAL BACKGROUND

When George Wolf was a graduate student in psychology at Yale in the early 1960s, Neal Miller offered him a choice of topics to study. Miller's interest was primarily in learning and motivation, and in developing simple experimental systems for the study of psychological phenomena. By this time, Miller had embarked, with his students, on the study of how the brain produces motivated innate and learned behaviors, including salt ingestive behavior (see his collected works, 1971a, b). George was given the choice to work on the problem of visceral learning, or the hunger for sodium. He chose the latter. But, when the time came to defend his thesis, the following question emerged from members of his committee: 'what does the ingestion of salt have to do with psychology?' His defense was that the study of sodium hunger could serve as a system for the study of motivated behaviors and how the brain generates them. To appreciate this, the study of sodium hunger does not begin with George Wolf, but with Curt Richter.

It was Curt Richter who discovered the phenomenon of sodium hunger in 1936. Several years later (1939) he suggested that the hunger for sodium was an innate drive. But, this was left to others to demonstrate. For example, Eliot Stellar, interested in how the brain produces motivated behavior, with his student Alan Epstein, showed that the sodium-deficient rat ingests large quantities of sodium within a short period of time when the rat is exposed to it for the first time (1955). Many others inquirers found similar results (e.g. Denton, 1982; Wolf, 1969b).

The fact that the behavior is innate, and easy to study, allowed experimental psychologists to examine what the animal might learn about the location of salt and what it was associated with. That is, rats could demonstrate that they learned something about the significance of the salt when they were not sodium hungry and demonstrate what they learned when they were for the first time (Krieckhaus & Wolf, 1968; Krieckhaus, 1970). The hunger for sodium was therefore used to study how learning interacts with innate prewiring. This is what intrigued Miller & Stellar; here, sodium hunger could be used as a model system in which to study the interaction of learning and homeostatic regulation.

Richter (1941) discovered that hormonal signals also generate a sodium hunger. In fact, hormones essential to the physiological regulation of sodium balance (angiotensin and aldosterone) also generate the behavior of salt ingestion (e.g. Fluharty & Epstein, 1983). The synergy of angiotensin and aldosterone is important to the arousal of sodium hunger (e.g. Epstein, 1984). However, other hormones also generate a sodium hunger. The hormones of female sexual reproduction (prolactin, oxytocin, estrogen), in addition to the hormones that are involved in the adaptation to stress (ACTH and the glucocorticoids), also generate a sodium hunger (Denton, 1982).

It was Richter who also first found that, during reproduction, the female rat's ingestion of salt goes up markedly (Richter, 1956; Denton, 1982). Salt intake is also sexually dimorphic. Females ingest two to five times more salt than male rats under a variety of conditions (Krecek, 1973; Wolf, 1982). Perhaps the greater hunger for sodium in females may have evolved because of the needs of the female during pregnancy and lactation.

A number of behaviors are sexually dimorphic and under hormonal control (Goy & McEwen, 1977). It is more generally known that steroid hormones, during critical periods of development, have profound effects on brain and behavior. Salt ingestion is one of them; virgin female rats typically ingest more salt than male rats, and the elevated salt ingestion, normally seen in adult female rats, is altered by manipulation of gonadal hormones during the neonatal period of development (Krecek et al., 1975).

These steroid effects are manifested because steroid hormones have both organizational and activational effects on both brain and behavior (Goy & McEwen, 1977). Neural circuits are shaped, and behavior is directed, by hormonal influences. Moreover, changes in the structure of the brain that result from steroid hormone actions are not restricted to just critical stages in development (Arnold & Breedlove, 1985). This is also true in the study of sodium hunger since the intake of salt is influenced by the animal's history during adulthood (e.g. Falk, 1965b). The hormones of sodium homeostasis, aldosterone and angiotensin, when elevated, also result in the increased responsiveness to salt even when rats no longer hunger for it (Sakai *et al.*, 1987, 1989).

Probably the best-studied hormone-induced behavior is lordosis in female rats. Estrogen and progesterone play an essential role in this female sexual behavior, and the neural circuit that mediates that behavioral effect has been successfully analyzed (e.g. Pfaff, 1980; McEwen *et al.*, 1987). For example, tritium-labeled steroids were mapped to receptor sites in the brain by radioautography and other neuroscientific techniques. The behavior was aroused by the application of the hormone to critical brain regions (ventral–medial hypothalamus), and was suppressed by the inhibition of neurotransmitters being synthesized within this region. The anatomical connectivity of this region of the hypothalamus to other brain regions was worked out, with a resulting analysis of a steroid-induced behavior and its anatomical basis. There are other examples of steroid-induced behaviors whose anatomical bases are now beginning to be understood (see Arnold & Breedlove, 1985). Now we add the study of sodium hunger.

We know something of where the brain hormone receptors that initiate sodium hunger are localized and of their mechanisms of action. In addition, the interaction of peripheral organs (heart, liver, etc), and the brain in the control of salt ingestion, have been outlined. Anatomical tracer studies have allowed for the assemblage of neural circuits that underlie salt ingestion. Thus, one begins to understand how the brain orchestrates the search for something salty by combining behavioral, physiological, anatomical, and biochemical techniques.

While in the search for something salty, animals use gustation to explore their world to select edibles from nonedibles. The question arises as to how the animal recognizes the salty taste, and which parts of the brain are involved. We are now in a position to address this. We know which gustatory (taste) nerves are importantly involved in the recognition of a salty taste, what the mechanisms of sodium transduction are, and at what level of the gustatory neural axis this occurs.

Richter thought that the change in sodium status resulted in an alteration of taste sensitivity and in changes in central states (see 1956 review). When this central state occurs, the animal then searches for, and ingests, the salt. While it is not the change that Richter suggested (increased sensitivity to salty ingesta), the specific need for sodium does, in fact, result in a change in taste sensitivity and in changes in central states.

This change in gustatory sensibility is an arbiter in the control of ingestive behavior. Sodium hunger is guided by specific taste fibers and neurons that are related to the salty taste of sodium. Significantly, the brain controls this behavior.

There is also a hedonic component to salt ingestion. That is what Pfaffmann called the 'pleasures of sensation' (1960), and it is a factor in taste-guided behavior. Richter and others (Nachman, 1962) certainly thought that salt tastes good to the animal that needs it. In fact, there is now strong evidence that gustatory sensibility and nutritional balance, particularly sodium balance, are, indeed, intimately connected, as both Richter and Pfaffmann had thought (e.g. Contreras, 1977).

But, it was P.T. Young (1949) who championed the idea that there is an hedonic factor in ingestive behavior. He did this at a time when such 'mentalistic' talk was sacrilegious, for he did his work in the heyday of American behaviorism. But, Young's criterion for the hedonic was always behavioral (choice and speed of approach), and he made an important distinction between palatability and appetite. Appetite is a response to what is needed – that is, sodium. Palatability responses can be independent of biological requirements.

As you will see, there is a dramatic shift in the oral–facial profile to infusions of hypertonic NaCl (about the concentration of sea water) when rats are sodium hungry. The facial display appears as if they are ingesting something rewarding – like something sweet – when they need the sodium (Berridge *et al.*, 1984). When they do not need sodium, they reject the salt in a characteristic fashion, and the hypertonic NaCl appears aversive. In fact, homeostatic needs like sodium hunger can produce affective changes in the acceptability of NaCl solutions. Similar phenomena have been described in other contexts, e.g. temperature regulation (Cabanac, 1971, 1979) and called 'alliesthesia', or the change in the hedonic response as a function of internal state (Stellar, 1980). In other words, to ensure motivation to ingest substances that normally are somewhat aversive, and to traverse distances and overcome obstacles to obtain the solute, Nature has tied the motivation to reward, and the reward includes pleasure when the salt is obtained.

For Curt Richter, behavior was essential in the regulation of the body's needs: the need for protein, carbohydrates, vitamins, and minerals such as sodium or calcium (see Richter's collected works, 1976; Schulkin, 1989). Richter brought to the study of behavior the idea of the regulation of the internal milieu (Bernard, 1865), and what was called homeostasis and the 'wisdom of the body' (Cannon, 1915, 1932).

The study of sodium hunger, therefore, was, and remains, an optimum place for the 'whole body' physiologist; that is, the physiologists with an interest in behavior were quick to see the importance of sodium hunger. Derek Denton is the outstanding exemplar. No one has studied salt ingestion to the extent and at as many levels as he has over the last 35 years. The experimental scope is vast. It goes from recording rabbits and kangaroos in the wild licking at salt pegs to the thorough analysis of the sodium hunger of sheep and the behavioral, physiological and anatomical mechanisms that control its behavior toward salt. The book he has written on sodium hunger is a classic (Denton, 1982).

The tradition of Denton is medical and physiological, with roots based on the works of Bernard, Cannon, and Richter. Their thesis is that behavior is organized physiologically by messages from the internal milieu to the brain. Perhaps, the messages emanate from sodium reservoirs (Wolf & Stricker, 1967; Stricker & Wolf, 1966), or from the activation of sodium sensors in the brain or liver, or from mechanisms in the heart, adrenals, kidney, and pituitary. The whole body figures in the regulation of body sodium. In other words, the study of sodium hunger allows one to examine the body at large and how it mobilizes itself to maintain sodium balance. One mechanism that it uses is behavior:salt ingestion.

In what follows, we will traverse a range of subjects. This will range from the ecology of salt ingestion to the sodium molecules and the action of various hormones. The emphasis is on experimental and behavioral manipulations of salt ingestion. The first chapter outlines the behavioral expression of salt-seeking behavior; the second, its hormonal regulation; the third is about the gustatory contribution; the fourth is the physiological, molecular, and pathological control or regulation; and the final chapter describes the neural circuits underlying salt ingestion. But, in each chapter, there is discussion of each – behavior, hormones, gustation, physiology, and the brain.

What the reader should walk away with is how a motivational system is organized at a number of levels, e.g. behavioral, physiological, and neurological. Reference is also made, throughout, to other motivations (e.g. hunger, thirst or sex), in addition to the study of affect, innate wiring, learning, and sexually dimorphic behaviors. The work has implications, more generally, beyond the study of sodium hunger.

1 Salt seeking behavior

SODIUM HUNGER: AN OCCURRENCE OF NATURE

The search for food and water which contain proteins, carbohydrates, vitamins and minerals is one of the major activities for most animal species. Evolution selects behavioral strategies that optimize the chances for satisfying bodily homeostatic demands. The behaviors include: 1) arousal; 2) search; 3) recognition; 4) decision to accept or reject; and 5) digestion. There are nutritional and sensory categories that help organize the behaviors (Rozin & Schulkin, 1990). Sodium hunger is, first, a phenomenon of Nature.

Periodically, rainfall depletes the sources of sodium that are being ingested from food by land-dwelling herbivores. This is especially true when the animal is located far from the sea, and probably contributed to the evolution of sodium hunger (Denton, 1982). The selection of a behavioral response to ingest salt is particularly true in females, perhaps because of the demands of reproduction (e.g. pregnancy and lactation – see Chapter 2).

A series of elegant experiments carried out in Australia (Blair-West *et al.*, 1967) demonstrated that, during times when the sources of sodium were low and diluted from rain, and the sodium-retaining hormone, aldosterone, was elevated, rabbits or kangaroos would travel to sodium sources placed by the experimenters to ingest this scarce commodity (Fig. 1.1).

In fact, sodium-deficient animals, particularly ungulates, are known to travel great distances to obtain and ingest salt (Denton, 1982). They are guided, perhaps, by a central gustatory system in addition to other sensory systems (e.g. olfaction and sight) that, like other sensory systems, reaches out to the world (e.g. Gibson, 1966). In this case, they are trying to find sources of sodium. The state of sodium hunger is one of exploration to find the needed sodium.

Thus, there are many instances in the wild of mammals searching for salt. Moose, white-tailed deer, reindeer, caribou, goats, sheep, and even elephants are known to search for salt deposits when sodium is scarce (e.g. Botkin *et al.*, 1973; Belovsky, 1978; Weeks & Kirkpatrick, 1976; Herbert & Cowan, 1971; Cowan & Brink, 1949). Such herbivorous animals tend to display energy maximizing strategies in the search for salt sources (Belovsky, 1978). Sodium in

Fig. 1.1. A telescopic lens photograph of a kangaroo at a salt peg (from Denton, 1982).

such contexts is just one of several minerals at the deposit. In fact, gorillas (Schaller, 1963) and chimpanzees (Goodall, 1986) are known to ingest dirt rich in many minerals. Not only do herbivorous and omnivorous animals ingest at such mineral licks when sodium is in short supply, but also, on occasion, carnivores are sighted at such licks (e.g. foxes). These carnivores are probably there to prey on herbivores.

The phenomenon is not limited just to mammals. It is also expressed in birds. For example, the vulture, red crossbill, and gold finch are known to eat at salt licks (Aldrich, 1939; Coleman, Fraser & Pringle, 1985; Peterson, 1942). This is particularly true during the reproductive season, in both birds and mammals, when females are known to be at the lick more often than males (e.g. Dixon, 1958).

In the laboratory, both sodium-hungry sheep (Denton, 1982) and sodium-hungry rats (Fig. 1.2; Arnell *et al.*, unpublished observations) will ingest salt licks. Therefore, sodium hunger is an occurrence of Nature, tied to the demands of evolution, and a phenomenon produced in the laboratory.

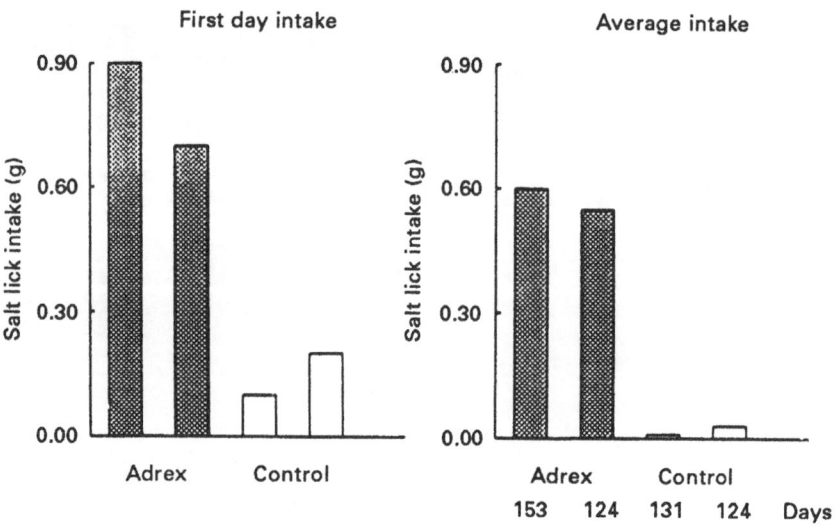

Fig. 1.2. Ingestion of salt lick in control and sodium hungry (adrenalectomized (Adrex)) rats (from Arnell, Schulkin & Stellar, unpublished observations).

INTRODUCTION TO EXPERIMENTAL DEMONSTRATION OF SODIUM HUNGER

What made sodium hunger attractive to study was that a rich set of behaviors were easy to elicit. In this chapter, I will describe some of the laboratory findings regarding the sodium–hungry animal, intermixed with reference to broader psychobiological issues. The animal I will be talking about mainly is the rat, but I will refer to the sheep, and, to a lesser extent, to the pigeon, the hamster, the mouse, the rabbit, and the monkey.

The reader should note, at the outset, that what is unique about sodium hunger, in contrast to vitamin or other mineral regulation, is that it is innate (see Rozin, 1976). That is, the first time rats or sheep are rendered sodium hungry, and then subsequently allowed access to salt, they ingest it within seconds before the body can appreciate the restorative consequences of the ingestion. Almost all vitamin or mineral hungers require some learning. In addition, rats, at least, do not have to be sodium hungry to learn about the significance of salt. In other words, they are prepared to learn where it is found, how to acquire it, and with what it is associated even in the absence of any motivation to consume salt. This should be a common occurrence in Nature; animals do not have to be in the relevant drive state, or in tissue deficit to notice a valued commodity and then return there when they need it (Krieckhaus, 1970).

The sodium–hungry animal is motivated to search for salt. It demonstrates what Wallace Craig (1918), the American naturalist, called both the appetitive and consummatory phases of instinctive behavior. In the appetitive phase, the animal searches for its goal, in this case a salty taste. In the consummatory phase, the animal has reached its goal and consumes the salt. Neither phase requires learning, though learning, no doubt, can influence both events.

Importantly, when an animal is sodium hungry there is a hedonic shift in the perception of salt (see also Chapter 3). In other words, the biological need for sodium is tied to the pleasure or reinforcement associated with salt ingestion. This, no doubt, contributes to the motivation of the animal to search for salt; that is, a hedonic judgment is made by the sodium-hungry rat that results in the salt now being perceived as more rewarding. Like other homeostatic systems, such as general food hunger or thermal regulation, the regulation is tied to hedonic events (Cabanac, 1971; Stellar, 1980).

Psychological, in addition to neurobiological, factors are discussed in this chapter, including: the enhancement of the avidity of the salty taste and its ingestion as a function of having been sodium hungry before, and the use of endogenous circadian clocks in the anticipation of salty commodities.

The picture I would like the reader to consider is the following: when an animal is sodium hungry, a representation of salty taste is activated, which serves to guide the behavior in the search, identification, and ingestion of salt. Innate mechanisms are responsible for the sodium-hungry animal ingesting the salt immediately upon its first exposure, and for noting the significance of the salt when it is not sodium hungry. A hedonic shift in the perception of the salt emerges in sodium hungry animals. The result is a motivated behavior with appetitive and consummatory phases in the search for a salty taste.

THE INNATE HUNGER FOR SODIUM

In the tradition of Cannon (1932), and his doctrine of 'the wisdom of the body', Richter envisioned a variety of innate behavioral capacities in order to select the nutrients and minerals that were needed. Others thought the behaviors were learned (e.g. Harris *et al.*, 1933). Richter was right about sodium hunger; he hypothesized early on that the appetite for salt which results from sodium deficiency is an innate drive (1939; 1956).

Richter's first study on salt appetite, over 50 years ago, used the adrenalectomized rat (1936). Because the adrenal gland had been removed, the animal was without the sodium-retaining hormone aldosterone, and chronically lost sodium. This is a fatal situation if there is no access to salt. With access, the adrenalectomized animal vigorously ingests sodium. The behavioral results are shown in Fig. 1.3. Richter also found that sodium salts were preferentially prefered over nonsodium salts (Fig. 1.4).

Richter did not provide evidence that the appetite is innate, he just hypothesized that it is. The sodium-hungry rat could have learned something

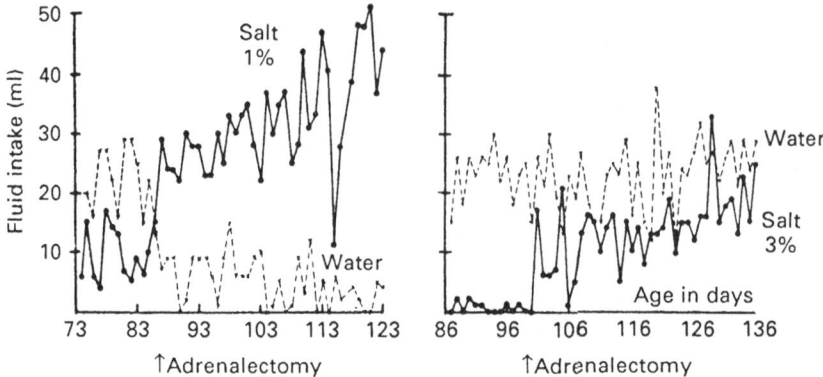

Fig. 1.3. 24-h salt and water intake before and following adrenalectomy (from Richter, 1936).

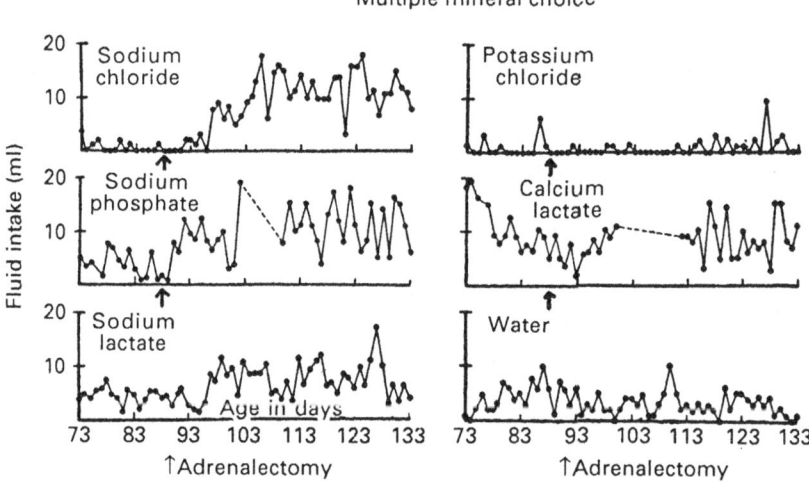

Fig. 1.4. 24-h salt intake before and following adrenalectomy (from Richter & Eckert, 1938).

from the consequences of the salt ingestion, or the exposure to the salt before the adrenalectomy. Bare (1949) also thought the hunger for sodium was innate. He showed that the adrenalectomized rat would ingest sodium within 24 hours. But, this finding does not show that the behavior is innate, since, by 24 hours, the rat may have learned something about the consequences of the salt ingestion. Are the drive and the recognition of the salt taste innate? Answers to these questions awaited further research.

Other inquirers (Epstein & Stellar, 1955) attempted to answer the question about the innateness of sodium hunger by tracing the development of salt appetite in adrenalectomized rats. In one study, one group of rats was immediately given access to salt postoperatively; another group was not given salt until they began to show the debilitating effects of chronic sodium loss. The ingestion test lasted for one hour. The interesting result was that both groups ingested comparable amounts. This suggests that sodium hunger may be innate, since it appears that the rats did not learn about salt by gradually ingesting it and restoring the sodium balance. Of course, it might be possible that, over the course of the one hour ingestion test, the rat may have learned something about the relationship between salt ingestion and the drive for sodium.

Sodium hunger was therefore demonstrated by showing that rats ingested the salt within 20 minutes. That is, the rats were sodium depleted for the first time and 24 hours later were given access to salty water (Falk & Herman, 1961). They responded vigorously to the sodium. Similar effects have been observed by others (Smith *et al.*, 1969). Further research (Nachman, 1962, 1963*b*) into the innateness of salt ingestion showed that adrenalectomized rats, or sodium-deprived rats, ingested the salt within seconds or minutes upon being exposed to it (Handal, 1965, Fig. 1.5). Sodium-depleted sheep also display an innate appetite for salt, despite no prior exposure to salt (Denton & Sabine, 1963). In fact, as with rats, the salt ingestion of sheep is rather rapid; they ingest most of the sodium within the first few minutes. (Denton, 1982). Therefore, they were not learning a relationship between the ingestion and the amelioration of the drive.

And, the fact that rat pups express the appetite is further evidence of its innateness. Rat pups can express competent feeding and drinking responses when removed from the dam (e.g. Leshem & Epstein, 1989). Salt appetite is also not just confined to the adult rat. Rat pups, as young as 12 days of age, which have never been exposed to sodium except what is contained in mothers' milk, and which were raised by a mother rich in sodium reserves, demonstrate a salt appetite the first time they are rendered sodium deficient (Moe, 1986). In the experiment, rats were adrenalectomized or acutely sodium depleted at 8 days of age. Four days later they were given access to NaCl solution. At the time of the test, the pups were taken away from the mother and then fitted with intraoral cannulae in which to infuse the NaCl. Body weight changes were the dependent measure. That is, body weight changes after the infusions were used as a measure of intake of fluids ingested. Figure 1.6 reveals that a hunger for sodium was expressed in the neonate rat as early as 12 days of age when sodium depleted.

In Nature, nutrients and minerals are found in food sources. Therefore, it should not be surprising that sodium hunger is not restricted to salty water (Bolles *et al.*, 1964). While an earlier study suggested that sodium-hungry rats

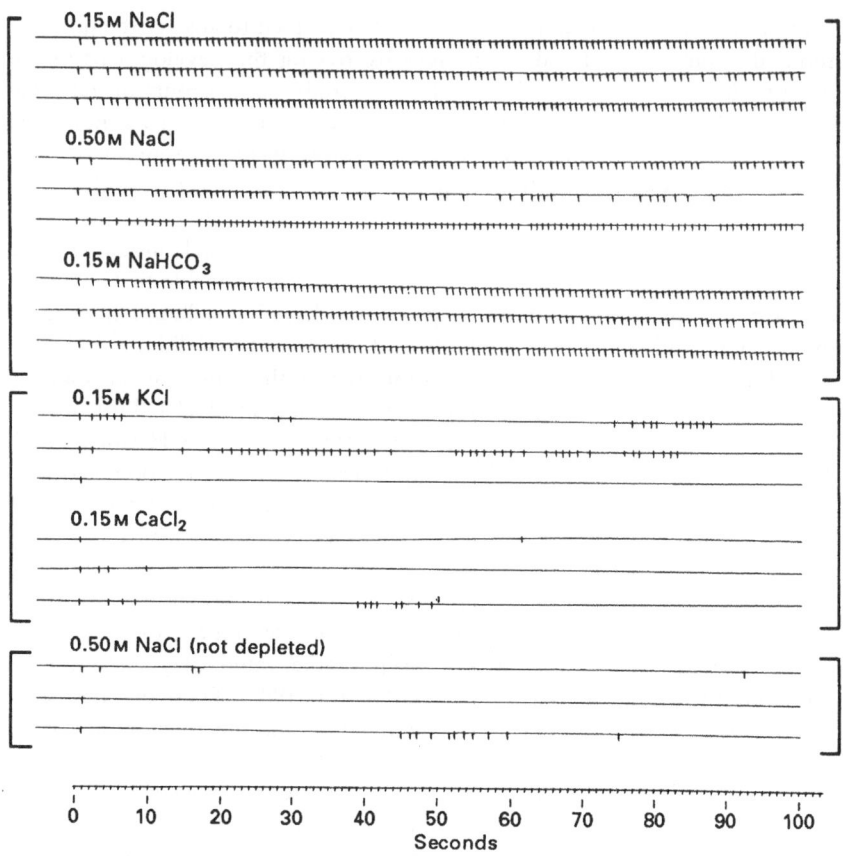

Fig. 1.5. Drinkometer records of individual rats. Every fifth lick is represented by a vertical mark. Records within the upper brackets are from sodium-depleted rats drinking solutions of sodium salts. Records within the middle brackets are from sodium-depleted rats drinking solutions of nonsodium salts. Records within the lower brackets are from nondepleted rats drinking a sodium salt solution (from Handal, 1965*b*; Wolf, 1969*b*).

would not regulate salt by ingesting salt in food (Fregly *et al.*, 1965), subsequent studies demonstrated that they would (Grimsley, 1968, 1970). Sodium hungry rats do, however, prefer to drink salty water, but, when salty water is not available, they will ingest salty food in greater amounts (Bertino & Tordoff, 1988). Furthermore, this response is also unlearned (Bolles *et al.*, 1964). In another study, using salty food, it was demonstrated that sodium-hungry rats, offered choices between salty and nonsalty foods, ingest the salty one (Rodgers,

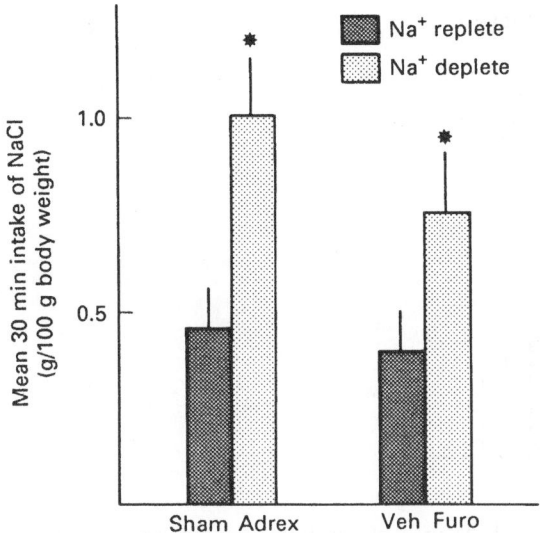

Fig. 1.6. NaCl intake (g/100 g body weight) of sodium-depleted 12 day-old rat pups. *Left*: Consumption of 6% NaCl by adrenalectomized (Adrex) or sham-operated (Sham) pups. *Right*: Consumption of 8% NaCl by pups injected with furosemide (Furo) (to produce Na depletion) or its vehicle (Veh) (from Moe, 1986).

1967). Young rats placed on a sodium-deficient diet for 21 days, and then given access to their regular food which was made salty and the choice of a novel food, ingested the salty food. This is in contrast to other mineral and vitamin-deficient rats who prefer a novel food over their old food which was providing the required vitamin or mineral (Rodgers, 1967; Rozin, 1976; Chapter 3). These mineral and vitamin appetites are learned, so the strategy is to ingest something different and then determine the consequences.

Taken together, the results reveal that the salt appetite that results from body sodium deficiency is truly an innate hunger but can be influenced by learning. Now we turn directly to how learning impacts on this innate hunger.

INFLUENCE OF LEARNING

It seems likely that, in Nature, it is not uncommon for animals to note significant objects in their environment, despite the fact that the objects may be irrelevant to their present concerns. Their future survival depends upon such learning. In the language of experimental psychology, this is called 'latent learning'. Tolman (1949), for example, provided some evidence for the phenomenon. But, for Hull and many other experimental psychologists, learning was thought to be dependent upon the reduction of a drive (1943). An

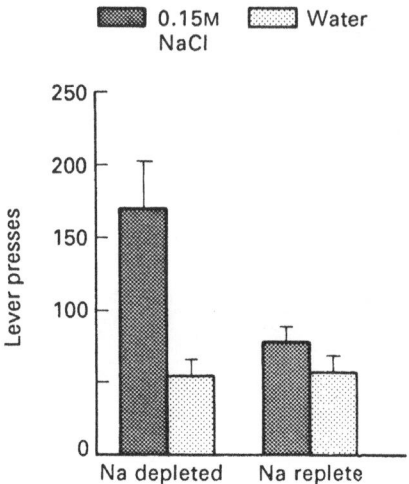

Fig. 1.7. Mean (SEM) number of lever press of rats during a 1 h extinction test. Rats either tasted NaCl or water during the training, and are now sodium depleted or sodium replete (from Krieckhaus & Wolf, 1968).

animal learned about its surroundings when it was hungry or thirsty; otherwise, it did not learn. Sodium hunger was also used in the analysis of this psychological issue.

The experimental design for the study of latent learning is elegant. Consider it for a moment. In one of the original experiments, thirsty rats were trained to bar press, half of them for water and half for isotonic saline (Krieckhaus & Wolf, 1968). Then they were rendered sodium hungry and given back their water to drink *ad libitum* in their home cage. The next day, the rats were once again put in the Skinner box; but the conditions had changed: there was no reinforcement, that is, no sodium was delivered to the animal following its pressing the bar for salt. Resistance to extinction was the measure of performance: how much the bar was pressed despite the fact that the sodium was not delivered. Only those sodium-hungry rats that ingested the saline when thirsty show pronounced bar pressing (Fig. 1.7). This phenomenon has been replicated many times (e.g. Morrison, 1971; Weisinger *et al.*, 1970). It has also been shown that other drives, such as hunger or thirst, do not elicit this response after the rat has tasted NaCl (Khalil & Eisman, 1971; Dickinson & Nicholas, 1983*b*). The response is thus specific to the sodium hunger motivation. In addition, the rats retained the propensity to bar press when subsequently rendered sodium hungry up to 8 days later (Krieckhaus, unpublished observations; Dickinson & Nicholas, 1983*b*).

Interestingly, if the rat is rendered sodium hungry one day following the training with different solutes, only the rats that tasted sodium will bar press for that salt (Krieckhaus & Wolf, 1968; also see Morrison, 1971). However, if one waits eight days between the training and the test, the rats bar press for the somewhat salty tasting KCl nearly as much as for the NaCl (Dickinson & Nicholas, 1983a). As I will elaborate further in the chapter on gustation and salt appetite (Chapter 3), the taste of KCl is chalky and bitter, as well as salty to humans. I speculate that, initially, the chalky taste overrides the salty taste, and the animal does not bar press for it. Over time, the chalky and bitter taste drops out of memory and the animal only recalls the salty taste, which is more prominent. That is why the sodium-hungry rat bar presses for the KCl. After all, the word salience has its roots in the word saline – the saltiness of time (Shakespeare).

Contextual cues play an important role in learning (Dickinson, 1980; Rescorla, 1981). In particular, it appears that contextual cues contribute to whether the rat will demonstrate this response of bar pressing for salt under extinction conditions (Dickinson, 1986). That is, this operant response appears neither to be related to how the salt was delivered during the animals' training with the reinforcer nor to whether the salt was delivered contigently, or noncontigently. The rats trained with NaCl bar press, none the less, when sodium deprived (Dickinson & Nicholas, 1983a,b). In the experiment, rats were trained to either receive saline or water from performing two operants. When sodium hungry, they did either behavior: chain pulling or bar pressing. There was no relationship between whether the salt was made available for the one performance and not for the other during training. Therefore, when the animal was made sodium hungry, it remembered the environment in which salt was made available, particularly where it was located, and not necessarily a specific operant that it performed in acquiring it.

Innate behavioral systems often require minimal exposure to the relevant stimuli to trigger their expression (Marler & Hamilton, 1966). A good example of this is bird song (Nottebohm & Arnold, 1976). Recognizing the significance of salt is another example. The rat also needs only minimal salt taste experience for the latent learning effects described above. Just five to ten NaCl licks is enough to elicit this response. That is, rats were first trained with water. Then, they were given five to ten licks that amounted to less than 0.5 ml of NaCl. Next, they were rendered sodium hungry, and given back their water in their home cage. The next day, they were again placed back in the Skinner box (Wirsig & Grill, 1982; Bregar et al., 1983). The results from one study are shown in Fig. 1.8. The group exposed to the NaCl for the brief period demonstrated the basic phenomenon of salt hunger described above. In addition, removal of the neocortex does not interfere with this phenomenon (Wirsig & Grill, 1982). Moreover, they retain this knowledge even if they learned the task before the

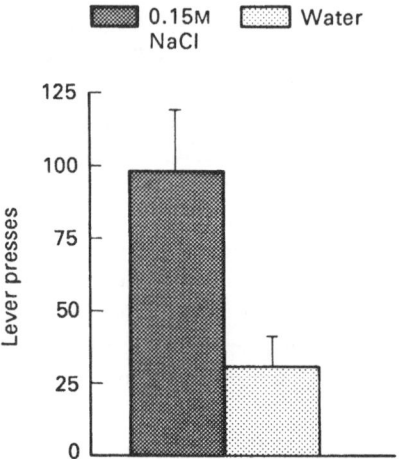

Fig. 1.8. Mean (SEM) number of lever presses of rats during a 1 h extinction test. Sodium-depleted rats either tasted NaCl or water during the training (from Bregar, Strombakis, Allan & Schulkin, 1983).

neodecortication. This suggests that the learning and storage of this capacity is beneath the neocortex.

Further corroboration, that minimal exposure to sodium produces a latent learning phenomenon, comes from an ablation study. In this study, damage to the central gustatory system at the level of the caudal thalamic relay, disconnecting afferent and efferent pathways into the ventral forebrain, impairs salt appetite (Wolf, 1968*b*). But, if the animals taste salt preoperatively and the salt is kept in the same place postoperatively, the rats with central gustatory damage ingest the sodium when they are rendered sodium hungry (Ahern *et al.*, 1978; Hartzell *et al.*, 1985). In Hartzell *et al.*, rats were given access to salty water or sucrose for 30 seconds while preoperatively thirsty. Only the rats that ingested the salt preoperatively demonstrated salt appetite postoperatively.

Animals recognize and remember where things are in space. Place learning is an important phenomenon (e.g. Tolman, 1949). The experiments on place learning for salt corroborate this fact. As I indicated, it is a probable occurrence of Nature that many animals encode the whereabouts of important sources of nourishment when they are not hungry or thirsty. Later, when they are hungry, they return to the place where the sources of nourishments were located. For sodium hunger, this phenomenon was demonstrated in an experiment by Krieckhaus (1970). He trained rats when they were thirsty to run a T-maze, where they either received water or a NaCl solution. When subsequently rendered sodium hungry, they preferred the side of the maze where NaCl had been formerly available. There was no such tendency when the rats were sodium replete. Data from central gustatory ablation studies provide additional evidence for place learning. If rats taste salt before the ablation, they ingest the salt postoperatively (Ahern *et al.*, 1978). One hypothesis is that, despite the diminished response to salt stimuli due to brain damage, the rat remembers where it tasted the salt preoperatively and returns to the place when rendered salt hungry postoperatively. This hypothesis was corroborated (Paulus *et al.*, 1984). Rats were trained to ingest salt or water in particular places in their cages preoperatively. Then, postoperatively, the salt position was switched for one group and kept the same for another group. The positions of the solutions are depicted in Fig. 1.9. The results, as shown in Fig. 1.10, reveal that the group that tasted salt preoperatively and postoperatively in the same place demonstrated salt appetite. The group that tasted the salt in a different position, pre- and postoperatively, were not protected against the effects of the lesion. Place learning, therefore, seems to play an important role in whether the rat with central gustatory damage will ingest salt. Taken together with the other results, it shows that rats with central gustatory damage manifest the same degree of postoperative protection of salt appetite after tasting salt for 4 weeks, 1 week, or 30 seconds. That is, once triggered, the salty engram is in place.

Time is another psychobiological category fundamental in animal learning (Rosenwasser & Adler, 1986). Endogenous biological clocks, particularly the circadian clock, organize behavior (Richter, 1965). Richter, in fact, demonstrated the importance of biological clocks and their role in the organization of behavior. Many forms of foraging behavior are determined by these clocks. Anticipatory behavior is known to depend upon a circadian rhythm, which reflects a specialized clock–dependent learning between availability of food and the time of day (Rosenwasser & Adler, 1986). We wanted to discern whether this held for sodium hunger.

First, note that rats are known to ingest salt as they do food and water during the dark phase of the light–dark cycle (Rowland *et al.*, 1985; Ikonomov *et al.*, 1985). This is not terribly surprising since this is the time the animal is awake. What is surprising is that, when rats were asked to show anticipatory running in a wheel, dependent upon a 24-hour timing clock to salt as they do for food, they did not do it for the salt under a variety of natriorexigenic conditions (Rosenwasser *et al.*, 1985). Since anticipatory running behavior may be inherently related to foraging behavior for food, we decided to test another behavior that had revealed circadian control of anticipatory behavior to food.

Rats show a variety of anticipatory behaviors, wheel running and lever pressing, before the availability of food (e.g. Rosenwasser & Adler, 1986). Sodium-hungry rats (adrenalectomized) were taught to bar press for salt (Rosenwasser *et al.*, 1988). Under *ab libitum* conditions, the bulk of the bar pressing for NaCl is clearly during the dark phase of the light–dark cycle. As the

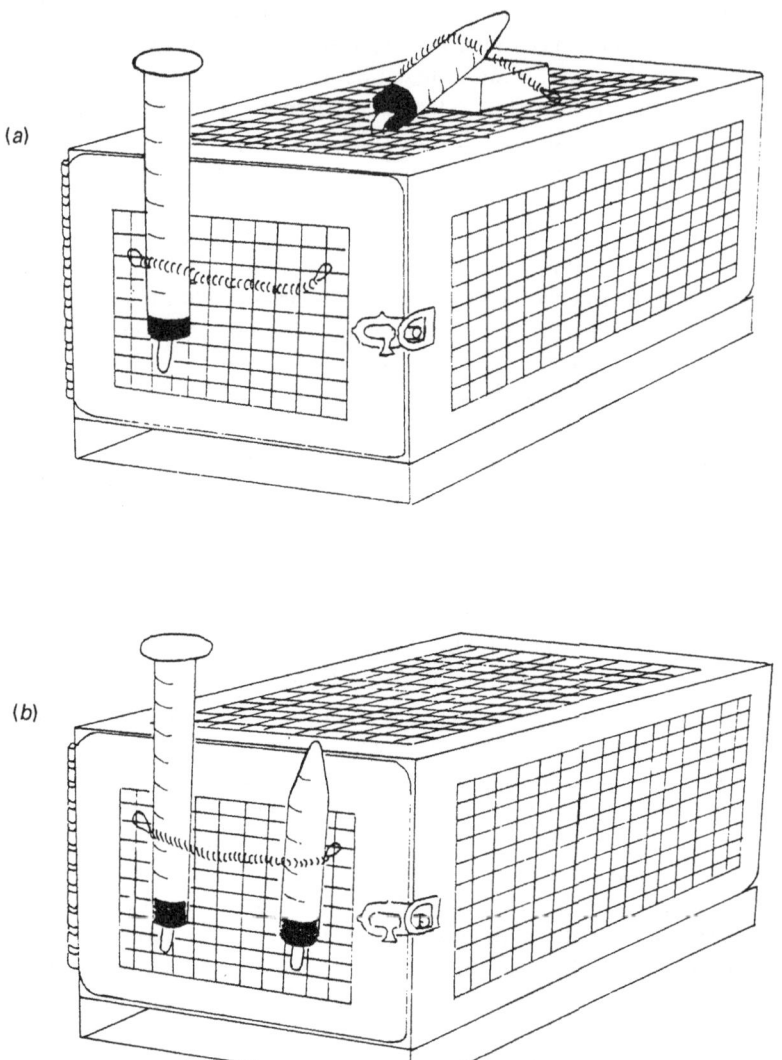

Fig. 1.9. (*a*) Position of 0.5M NaCl and water tubes during preoperative experience for the same-place group. (This is also the position of two water tubes during preoperative experience for the water control group and the position of salt and water tubes during postoperative salt appetite test for all groups.) (*b*) Position of salt and water tubes during preoperative experience for the different-place groups (from Paulus, Eng & Schulkin, 1984).

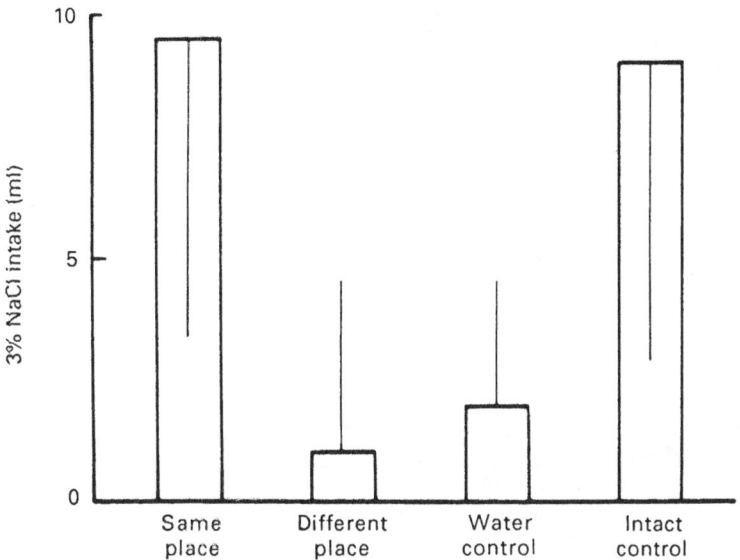

Fig. 1.10. Median increases of NaCl intake over baseline intake, with the interquartile range indicated (from Paulus, Eng & Schulkin, 1984).

salt availability becomes restricted to 2 h per day, rats demonstrate anticipatory operant behavior (Fig. 1.11). Therefore, sodium-hungry rats are able to anticipate salt availability without external cues (that is, someone in the room about to give the salt, or a tone, etc) by using endogenous circadian clocks. Note that salt availability at other times did not provoke this anticipatory behavior. But, also note that, though we were able to show some anticipatory behavior, the phenomenon in the sodium-hungry rats is less pronounced than in food-deprived rats (Rosenwasser & Adler, 1986). It is also difficult to get thirsty rats to show anticipatory behavior for water that is dependent upon an endogenous circadian timing mechanism (Mistlberger & Rechtschaffen, 1985). The fact that sources of sodium and water may be more stationary than sources of food may have something to do with this. That is, water and mineral licks tend to be more localized, perhaps, than sources of food; therefore, the circadian clock may be inherently related to the timing of food sources that appear at different locals on a daily basis.

Rats, and other animals, associate different taste properties together (e.g. salty, chalky, sour). For example, rats that were made sick from ingesting a compound taste stimulus reject the individual components when they are subsequently given alone (Rescorla, 1981), or they come to prefer specific tastes because of their associations with other tastes (Rozin & Zellner, 1985). In

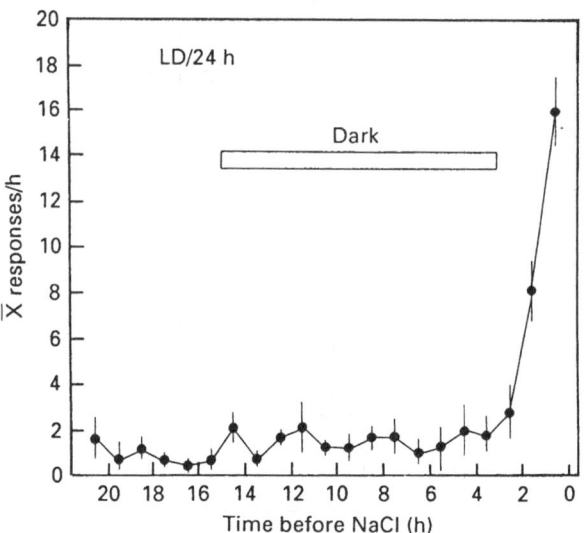

Fig. 1.11. Mean number of nonreinforced lever presses per hour under salt-access schedules. 'Light–dark 12:12 h and 3 h of salt access. (from Rosenwasser, Schulkin & Adler, 1988).

Nature, salty tastes are associated with other taste properties at salt licks. And, in the laboratory, rats are also prepared to associate the taste of salt with other tastes and then to ingest them subsequently when rendered sodium hungry. Fudim (1978) provided the first evidence for this. He trained rats to ingest salt mixed with banana or almond. When rendered sodium hungry the first time, the rats ingested the flavor that was associated with the sodium; that is, they ingested the flavor that had been associated with the NaCl. For an interesting discussion of such sensory preconditioning, see Rescorla (1981).

This work was extended in several ways. First, it was shown that the phenomenon could be demonstrated with quinine or citric acid, which are less palatable solutions than almond or banana used in the Fudim study (Rescorla, 1981; Berridge & Schulkin, 1989; Fig. 1.12). Rats were deprived of water and trained to ingest the solutions within a 10-minute test period (Berridge & Schulkin, 1989). Then they were given the quinine or citric acid alone. Rats, exposed to the compound stimulus, show the sensory preconditioning effect.

It seems clear that rats are prepared to associate the location of salt, in space and its appearance in time, and what it is associated with when they are sodium replete. Perhaps, similar phenomena would hold for other innate behavioral regulatory systems. For example, since there is some evidence that rats innately recognize specific proteins (Deutsch *et al.*, 1989), and that chickens innately

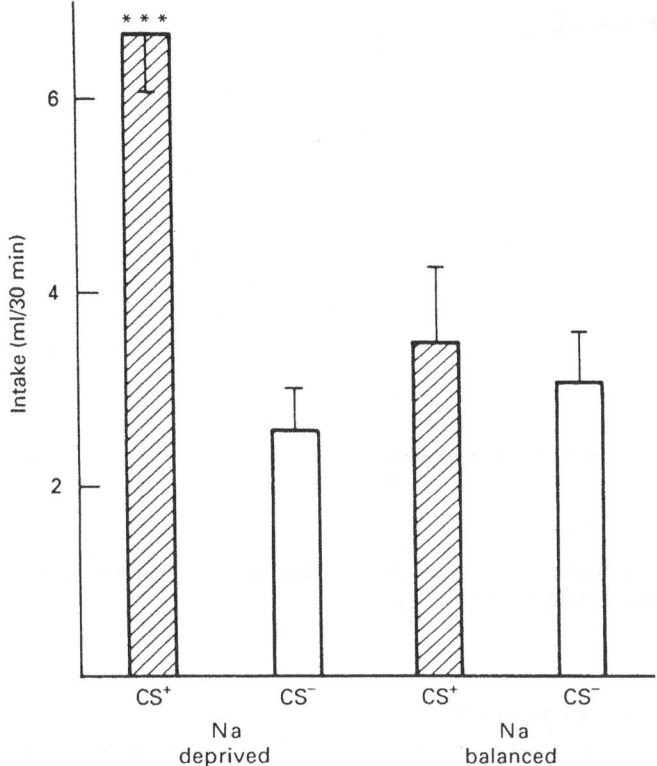

Fig. 1.12. Mean (SEM) amounts consumed of the CS (quinine or citric acid) and CS solutions in a 30-min intake test while either sodium balanced or when sodium depleted (from Berridge & Schulkin, 1989).

recognize water (Stricker & Sterritt, 1967), such latent learning phenomenon should be demonstrable in these other systems.

Finally, social factors also play an important role in food and mineral ingestion (e.g. Galef, 1986; Rozin & Schulkin, 1990). Food preferences are, in part, determined by watching other conspecifics, and what they eat (Galef, 1986). The ingestion of salty food is, to some extent, modified by such learning. Rats that have observed conspecifics, eating food with varying sodium content, will tend to eat the salted food they observed their conspecifics ingesting, independent of the sodium content when they are sodium hungry (Galef, 1986). That is, observing what others have eaten can have consequences for the sodium-hungry rat.

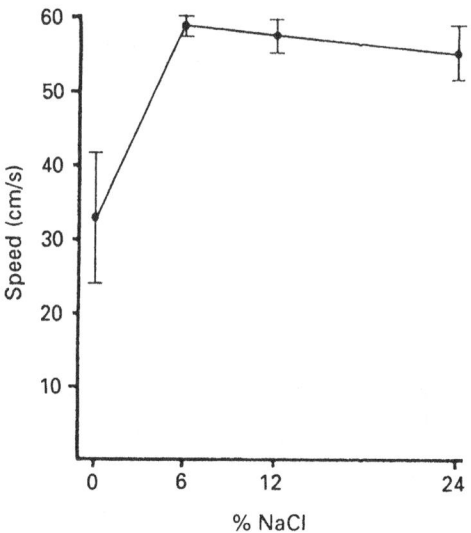

Fig. 1.13. Running speed for NaCl when rats are sodium hungry, generated from DOCA (from Schulkin, Arnell & Stellar, 1985).

MOTIVATION TO INGEST SALT

As I indicated, there is an appetitive phase followed by a consummatory phase in motivated ingestive behavior (Craig, 1918). Expectancy plays a large role in motivated animals, and motivated animals are usually able to perform operant behaviors to obtain what they need. This has been nicely demonstrated when rats are hungry, thirsty, or when sexually motivated (Stellar & Stellar, 1985). It holds for the sodium-hungry animal.

Consider the appetitive phase. The sodium-hungry animal is motivated and willing to labor to obtain something salty. Sodium-depleted animals easily learn to bar press for salt (e.g. Wilson & Wong, 1975). The phenomenon, for example, has been demonstrated in rats (Lewis, 1960), sheep (Weisinger *et al.*, 1983) and calves (Bell & Sly, 1979). All three species are willing to expend the energy to bar press for salt when under states of sodium depletion. In both rats and sheep, the greater the sodium hunger, the greater the amount of salt that is ingested (see Denton, 1982); therefore, the degree of drive in part determines the ingestion. A very interesting experiment by Quartermain & Wolf (1967) also showed that rats that were sodium depleted for the first time bar pressed for salt depending upon the degree of the sodium treatment; the greater the sodium depletion, the more the animal bar pressed for salt. They would not, however, bar press when the reinforcement was water instead of salt. Similar results were

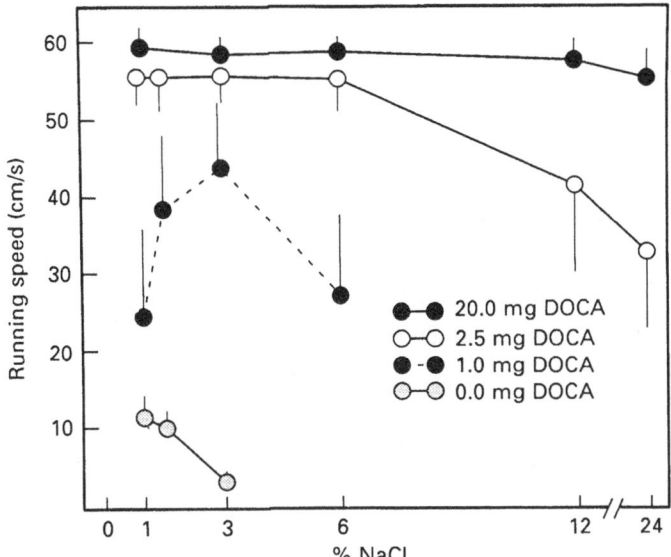

Fig. 1.14. Running speed for NaCl under different degrees of sodium hunger, produce by different doses of (deoxycorticosterone) DOCA (from Schulkin, Arnell & Stellar, 1985).

obtained in sodium-depleted sheep (see Denton's review, 1982) and in mineralocorticoid-treated rats (see Chapter 2, Wolf & Quartermain, 1966). Therefore, the response is not just reflexive (Wolf, 1969*b*).

The appetitive phase of motivation was also demonstrated in rats that run for salt when they are rendered sodium hungry (mineralocorticoid treatment coupled with being placed on a sodium-deficient diet; Schulkin *et al.*, 1985). The running speed of the rat was measured and it was given a small draft of NaCl at the end of the runway. The first time they were given this treatment, they ran faster for salt, than for water (Fig. 1.13). The study also demonstrated that running speed was related to the degree of the drive and the concentration of the taste of salt. One can see from Fig. 1.14 that the greater the sodium hunger the faster the animal will run for the salt, and the more concentrated the salt the more likely the animal will decrease its running speed.

HEDONIC CHANGES IN THE PERCEPTION OF SALT

Having discussed the appetitive phase, now consider the consummatory phase. In humans, there are characteristic facial displays to bitter and sweet solutions in the neonate, in the ancephalic child and in the normal adult (Steiner, 1977; Nowlis, 1977). Facial displays are one indicator of affect (Darwin, 1872), and the hedonic appreciation of ingesta. The analysis of human facial displays has also

Ingestion
sequence

Aversion
sequence

Fig. 1.15. Taste-elicited consummatory responses. Ingestive responses (*top*) are elicited by continuous oral infusions of glucose, sucrose, and other palatable solutions and include rhythmic mouth movements, tongue protrusions, lateral tongue protusions, and paw licking. Aversive responses (*bottom*) are elicited by infusions of quinine solutions and include gapes, chin rubs, head shakes, face washing, forelimb flailing, and locomotion (not shown) (from Grill & Norgren, 1978*a*).

been extended to rats, which reveal characteristic facial responses to ingesta similar to those of humans; when they are infused with something sweet, they express a positive ingestive oral–facial response; when infused with something bitter, they express a negative rejection oral–facial response (e.g. Grill & Norgren, 1978*a*; Berridge & Grill, 1983). Taste aversion learning (Garcia *et al.*, 1974) has been used to show that a sweet taste can be rendered aversive by poisoning (Berridge & Grill, 1983). And, a concentrated NaCl solution can be rendered positive by sodium hunger. In other words, the analysis of the oral–facial response was applied to the ingestion of NaCl to test the hypothesis that sodium-hungry animals enjoy the taste of salt.

The taste reactivity method developed by Grill & Norgren (1978*a*) was used to test this idea directly. In this context, rats are infused with taste solutions through intraoral cannulae, and their facial responses are videotaped. Figure 1.15 depicts species typical oral–ingestive behaviors. There are characteristic ingestive and rejection responses to tastants (see also Chapter 3). When rats are sodium replete, and hypertonic salt is infused into their oral cavity, their ingestive responses are minimal and their aversive responses are mixed. When they are sodium hungry, the pattern changes dramatically. The concentrated hypertonic NaCl is now avidly ingested (e.g. Berridge *et al.*, 1984; Berridge & Schulkin, 1989; Fig. 1.16). There are few or no aversive responses and ingestion increases; they show pure ingestive responses. This suggests that there is a hedonic shift in the perception of salt when the rats are sodium depleted. In this context, this hedonic shift is unlearned. Sodium-hungry rats never exposed to the taste of salt show the same ingestive response, while the controls do not (Berridge *et al.*, 1984). Moreover, they do not show this shift to equally aversive nonsalty tastants such as citric acid (Berridge *et al.*, 1984).

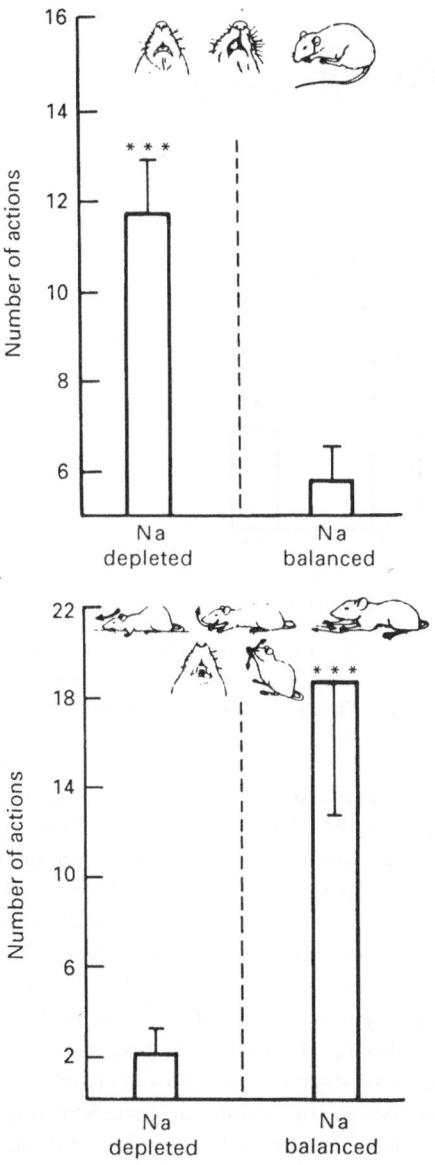

Fig. 1.16. Taste reactivity profile to 0.5M NaCl while either sodium balanced or when sodium depleted. *Top*: combined mean (SEM) number of ingestive actions (rhythmic tongue protrusions, nonrhythmic lateral tongue protrusions and paw licks). *Bottom*: combined mean (SEM) number of aversive actions (chin rubs, head shakes, paw treads, gapes, face washes and forelimb flails), (from Berridge & Schulkin, 1989).

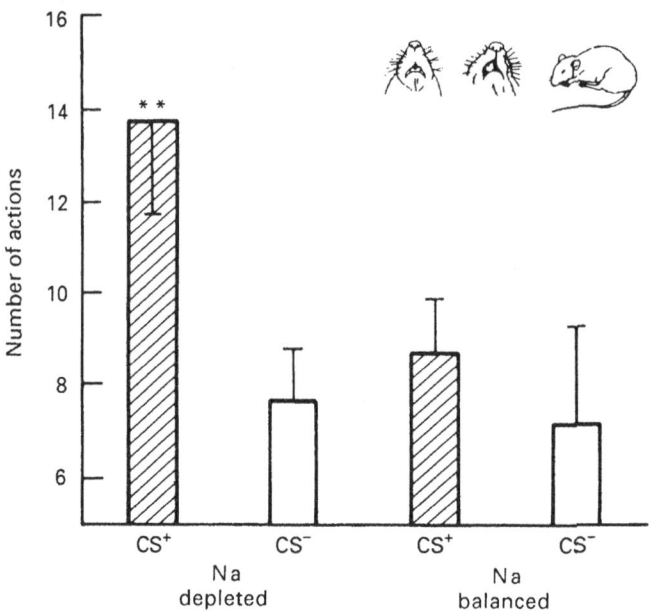

Fig. 1.17. Ingestive taste reactivity to conditioned tastes. Mean (SEM) number of combined ingestive actions emitted to the CS (quinine or citric acid) and CS while either sodium depleted or when sodium replete (from Berridge & Schulkin, 1989).

The sensory preconditioning experiments, described earlier, have also demonstrated a hedonic shift in taste qualities associated with sodium. That is, it was demonstrated that an arbitary taste associated with sodium becomes hedonically more palatable to the sodium-hungry animal (Berridge & Schulkin, 1989). The design was simple. Rats were intraorally infused through oral fistula; then, during the sensory preconditioning phase of the experiment, rats were given mixtures of quinine or citric acid paired with sodium. This was repeated for several days for 10 minutes per day. Rats were then rendered sodium hungry for the first time. The following day their facial responses were videotaped. The group that had the quinine or citric acid and associated with sodium showed greater ingestive responses to the quinine or citric acid (Fig. 1.17). Interestingly, unlike the response when the sodium-hungry rat was infused with sodium, the negative facial display to quinine or citric acid was unchanged.

Taste qualities can be enhanced, and there are a variety of reasons why this occurs (for review, see Rozin & Schulkin, 1990). There are also various reasons why an animal will ingest salt, one factor is the rat's past experience of sodium depletion, or sodium hunger experiences. For example, Falk, 1966; Tang &

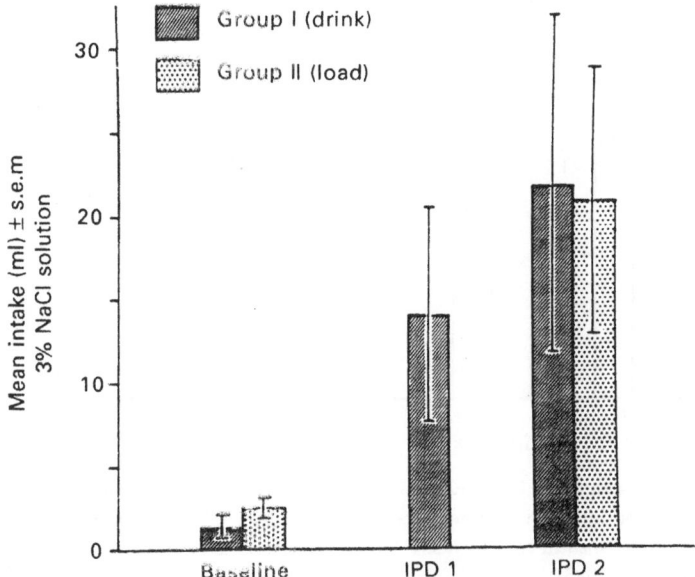

Fig. 1.18. Mean intake (ml) ±standard error of the mean of 3% NaCl solution during fluid test period (13 h) before (baseline) and after intraperitoneal dialysis (IPD). Group I drank NaCl solution after IPD 1, and IPD 2. Group II was stomach loaded with NaCl solution after IPD 2, and drank NaCl solution after IPD 2 (from Falk, 1966).

Falk, 1979 discovered that rats rendered sodium hungry ingested greater amounts of sodium on the second treatment than they did on the first (see also Berridge *et al.*, 1984). That is, rats depleted of sodium ingest greater amounts of salt by the second depletion (Sakai *et al.*, 1987; Berridge *et al.*, 1984). Prior experience of salt ingestion for this enhancement phenomenon also is not necessary (Fig. 1.18).

Interestingly, lateral hypothalamic damage impairs or abolishes the expression of sodium hunger (Wolf, 1964a, 1967; Ruger & Schulkin, 1980), as well as other ingestive behaviors (Epstein, 1971). But, if preoperatively rendered hungry or thirsty, they are protected against postoperative impairments in ingestive behavior (for review, see Schulkin, 1988). Similarly, rats treated once weekly with sodium hunger-inducing treatments are protected against the disruptive effects of the lesion on sodium hunger when tested postoperatively (Ruger & Schulkin, 1980). Salt intake is not necessary for this effect. That is, rats preoperatively rendered sodium hungry do not have to ingest the salt to be protected. They just need to have experienced the hunger for sodium. Other alimentary experiences such as hunger or thirst are not protective (Schulkin & Fluharty, 1985; Fig. 1.19). Perhaps this occurs because

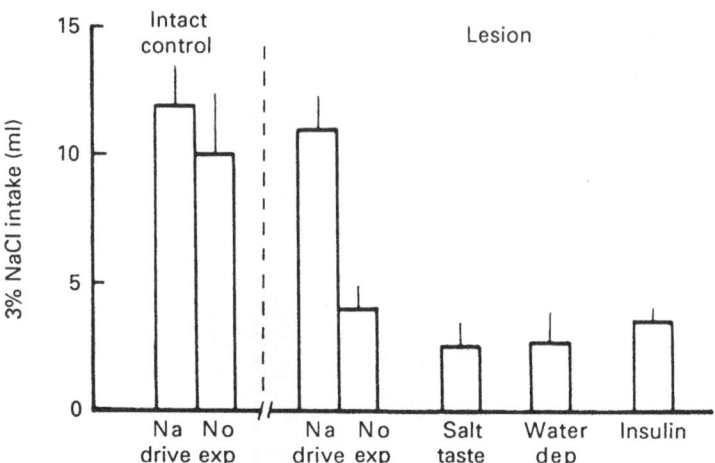

Fig. 1.19. Mean increased drinking over baseline conditions of NaCl in lesion rats given preoperative sodium drive experience, no experience, salt taste experience, water deprivation, or insulin treatments. (Vertical lines indicated SE) (from Schulkin & Fluharty, 1985).

the salt ingestion of the lateral hypothalamic rat has been elevated by the preoperative treatments.

In addition, need-free NaCl intake is elevated following these natriorexigenic treatments, despite the fact that the rats no longer need the sodium (Sakai *et al.*, 1989). This is also apparent in the lateral hypothalamic-damaged rat that was preoperatively rendered sodium hungry (Ruger & Schulkin, 1980). This phenomenon is particularly conspicuous in nonlesioned females (Fig. 1.20). Perhaps, the hormonal signals that result from the depletion produce long-term changes in brain function that alter the incentive value of the salt (Chapters 2 and 5).

OTHER EXPERIMENTAL EXAMPLES OF SODIUM HUNGER

Having discussed mainly the sodium hunger of the rat and the sheep, and its relationship to some extent with other ingestive or motivated behaviors, I would close this chapter by briefly considering the range of animals expressing this appetite by experimental manipulation. Recent attention by comparative psychobiologists have emphasized species differences in ingestive behavior; not all animals express a sodium hunger (Rowland, 1986; Rowland & Fregly, 1988*a*). Now consider the ones that do.

Birds demonstrate a sodium hunger. Pigeons are known to express an appetite for salt. Pigeons, that are sodium depleted, express an appetite for NaCl

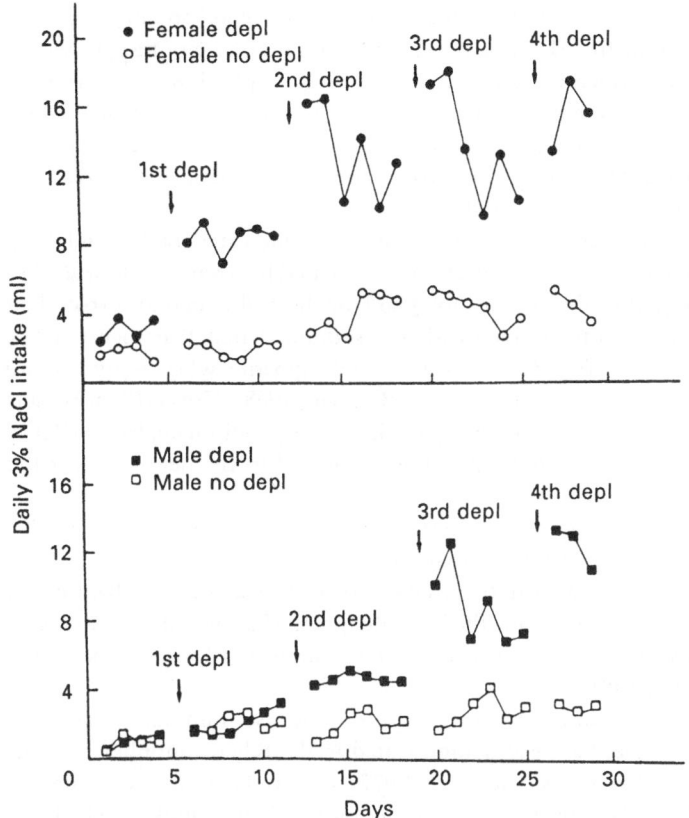

Fig. 1.20. Daily need-free intakes of 3% NaCl of depletion virgin (open) and depletion (filled) male (squares) and female (circles) rats for 30 days over the course of four successive sodium depletions (depl) (from Sakai *et al.*, 1989).

when exposed to it on the following day (Alan & Schulkin, unpublished observations; Epstein & Massi, 1987, Chapter 2).

In the rat, there are strain differences in the salt ingestion that results from body sodium depletion (see Chapter 3), but it is none the less expressed in a number of strains of rats (e.g. Grill & Bernstein, 1988). And while salt appetite has not always been elicited in mice (Rowland & Fregly, 1988*c*), they do, none the less, ingest sodium in response to body sodium depletion (Rowland & Fregly, 1988*c*; Denton *et al.*, 1988). When given a choice, they prefer sodium salts over other salts. They also demonstrate a moderate preference for NaCl when sodium replete (see Chapter 3) and, like rats, tend to overingest the sodium salt (Denton *et al.*, 1988; but see also Rowland & Fregly, 1988*a*). They not show more enhanced sodium ingestion on the second sodium depletion than they did on the first (Denton *et al.*, 1988), and they are not responsive to a number of natriorexigenic treatments (Rowland & Fregly, 1988*c*). Hamsters or gerbils do increase their salt intake when rendered sodium deficient (Cullen, 1972). It should be noted, however, that salt ingestion of the hamster is dramatically lower than it is for rats. None the less, the appetite for salt is clearly able to be elicited (Fitts *et al.*, 1983).

Ungulates also demonstrate a sodium hunger. As indicated, rabbits and sheep demonstrate salt appetite in the laboratory as elicited by a variety of means (see Denton, 1982). Unlike the rat, the rabbit and the sheep also tend to correct their sodium ingestion commensurate with the sodium deficit that incurred (see Chapter 3 for discussion). Goats express a salt appetite when body sodium depleted; they will also bar press for salt (Baldwin, 1968). Cows (Blair-West *et al.*, 1987), and calves, increase their salt intake to body sodium depletion (Bell & Sly, 1979). The degree of sodium hunger is related to the degree of salt loss (Blair-West *et al.*, 1987).

Even carnivores can express a sodium hunger. Carnivores are under less pressure to evolve a mechanism for salt ingestion, since they get their salt from their prey and do not appear to possess a special gustatory mechanism for detecting salt (Chapter 3). None the less, salt appetite has been demonstrated in the dog following sodium deprivation (Fitzsimons & Moore-Gillon, 1980; Ramsay & Reid, 1981).

Finally, monkeys and primates also demonstrate a sodium hunger. Sodium hunger, for example, has been demonstrated in the Rhesus monkey. Despite some doubts (McMurray & Snowdon, 1977), one study showed that they do (Schulkin *et al.*, 1984). Note that there is salt gluttony in the monkey, which like the rat, is an omnivore. Finally, in humans, McCance attempted to deplete himself and several other people of body sodium (1936, 1938). They sweated themselves, and ate a low sodium diet. The subjects felt weak and foods were generally perceived as less tasty or weaker. Bertino *et al.* (1981) also reported that young adults placed on a sodium-deficient diet also reported that salty ingesta were less intense, and Beauchamp *et al.* (1990) have reported that sodium-depleted humans perceive the salty taste as more pleasant (see Chapter 3).

Thus it should be clear that sodium hunger, induced in the laboratory, is a rather robust behavioral phenomenon. Moreover, the behavioral expressions are rich and diverse. While the behavior is innate, learning impacts to orient the sodium hungry animal. Hedonic changes motive; the result is the search for salt.

2 Hormonal regulation of salt intake

INTRODUCTORY

The hormonal control of behavior is an active area of research (Beach, 1971), and a variety of sexually dimorphic behaviors are hormonally induced, e.g. reproductive behavior (Phoenix *et al.*, 1959), bird song (Nottebohm & Arnold, 1976). Sodium hunger is hormonally controlled and a sexually dimorphic behavior, and, while not as sexy as other hormonally induced behaviors, it is revelatory about the hormonal control of behavior. Consider first, in this introduction, the background for mineralocorticoid-induced sodium hunger.

Hormonal events that regulate behavior and physiology evolved through natural selection, early on, as adaptive mechanisms for survival. The hormone aldosterone (ALDO), a mineralocorticoid, appears to have been essential in the move from the ocean to fresh water (Denton, 1965). Thus, the mineralocorticoids have come to be characterized as hormones for a 'tetrapod novelty' (Bern, 1967). ALDO played an important role in the colonization of land by allowing animals to retain sodium and, therefore, to maintain their sodium balance.

In some species, as you will see, ALDO came to play a role in generating the behavior of searching for, and then consuming, salty commodities. Salt ingestion is particularly enhanced during the reproductive season when there are greater demands on the female. The adrenal gland is enlarged and ALDO levels are elevated during this time; the adrenal gland (where ALDO is made) of females is larger presumably because females need to retain sodium for their offspring (see, e.g. Hoffman & Robinson, 1966).

The hormone ALDO serves primarily two functions. First, it contributes to the restoration of body sodium homeostasis during body fluid loss (discussed in Chapter 4) and during the demands of reproduction; second, it generates a hunger for sodium. This hunger for sodium may have evolved in females, where the need is greater and the avidity for salt is most pronounced in both the field and in the laboratory.

The hunger for sodium can be generated by a variety of hormonal signals in addition to ALDO, including both peptide and steroid hormones. The two principal natriorexigenic hormones are the mineralocorticoid ALDO and the peptide hormone angiotensin II. They act to maintain sodium levels and extracellular fluid volume. The principal antagonist of these hormones is atrial natriuretic factor, a peptide hormone found in both the atrium of the heart and the brain. It is an antagonist of the renin–angiotensin–aldosterone system physiologically and an antagonist of the renin–angiotensin system behaviorally. In addition, the major pituitary hormone, ACTH, generates a hunger for sodium along with the glucocorticoids that are related to stress or arousal and energy homeostasis. The female hormones of sexual reproduction also have a profound effect on salt intake. In particular, oxytocin, estrogen, progesterone, and prolactin generate an appetite for salt that is related to the female's need during pregnancy and which helps to restore the sodium that is required during gestation and then lost to her offspring during lactation.

PRINCIPAL NATRIOREXIGENIC HORMONES INVOLVED IN THE REGULATION OF SODIUM AND EXTRACELLULAR FLUID BALANCE

The hormones that defend against the loss of sodium in the body and that redistribute sodium during bouts of sodium deficiency also generate an appetite for salt. These hormones include angiotensin II and the principal mineralocorticoid hormones ALDO and deoxycorticosterone (DOCA). Nature has once again economized. It has provided these hormones of sodium homeostasis with behavioral as well as physiological actions. Consider angiotensin first.

ANGIOTENSIN-INDUCED SODIUM HUNGER

Angiotensin II is the hormone of 'extracellular fluid' regulation (Fitzsimons, 1979). It is known to elicit water drinking in fish, birds, and many mammals (Fitzsimons, 1979). It also plays a very important role in the regulation of body sodium homeostasis by generating an appetite for salt, and physiologically in the conservation of sodium (Chapter 4).

Braun-Menendez (1952) was the first to note that systemic renin injections elevate salt ingestion in rats. While others failed to find a role for systemically given renin or angiotensin in salt appetite (Fitzsimons & Stricker, 1971; Fitzsimons & Wirth, 1976), somewhat later it was clearly shown that systemic angiotensin did elevate salt ingestion in rats (Findlay & Epstein, 1980), and sheep (Weisinger *et al.*, 1987*a*, *b*). In the rat, these effects may be largely due to angiotensin-induced natriuresis (e.g. Fluharty & Manaker, 1983). Importantly, this possibility suggests that the appetite, in this instance, is not primarily due to angiotensin but to the loss of sodium; but then it is sodium loss that naturally generates angiotensin synthesis and secretion (see Chapter 4). But angiotensin should generate the appetite for salt in this experimental context independent of the sodium loss.

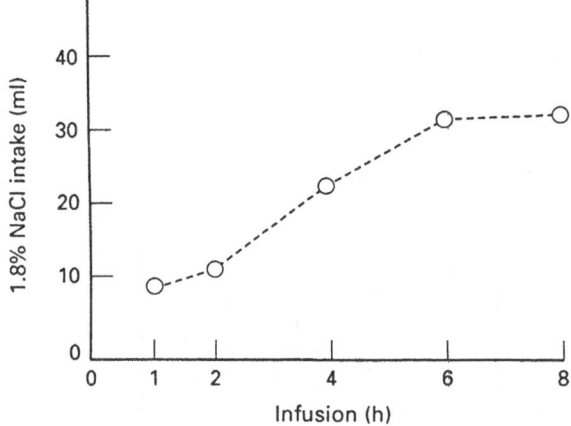

Fig. 2.1. Mean sodium intake during/after ventricular infusion of angiotensin at a rate of 50 ng/μl for 8 minutes (from Buggy & Fisher, 1974).

The finding that centrally administered angiotensin, delivered directly to the rat's brain, elevates water intake and dramatically increases salt intake suggests that this hormone indeed plays a dual role in body fluid homeostasis: an increase in water ingestion followed by an increase in salt ingestion (Buggy & Fisher, 1974). Figure 2.1 depicts the salt ingestion. These results have been corroborated by a number of laboratories (e.g. Wong & Whiteside, 1974; Chiaraviglio, 1976b). The appetite also appears following central renin injection into the lateral or third ventricles (e.g. Avrith & Fitzsimons, 1980a). In both instances, the appetite occurs before the onset of the natriuresis (DeLuca *et al.*, unpublished observations; Fluharty & Manaker, 1983). The appetite appears within an hour following the intracerebral ventricular injection; the natriuesis takes about 5 hours and is dose related.

There is also an enhanced avidity for NaCl following central renin–angiotensin injections. That is, there is a persistent appetite even after the angiotensin infusion is terminated (Bryant *et al.*, 1980). Finally, the avidity for salt in the adrenalectomized rat is further augmented following central angiotensin injection (Fitzsimons & Fuller, 1985). And, that centrally delivered angiotensin generates motivated behavior; rats will run down a runway for NaCl (Zhang *et al.*, 1984 described in Chapter 1).

While we do not know about the natriuretic actions of renin–angiotensin in the neonate, we do know, as indicated in Chapter 1, that sodium hunger is expressed in young rats. Renin–angiotensin-induced thirst, in addition to sodium hunger (e.g. Leshem & Epstein, 1989), is expressed in young rats (Fig. 2.2). These rats are taken from their mother, given a central injection of renin

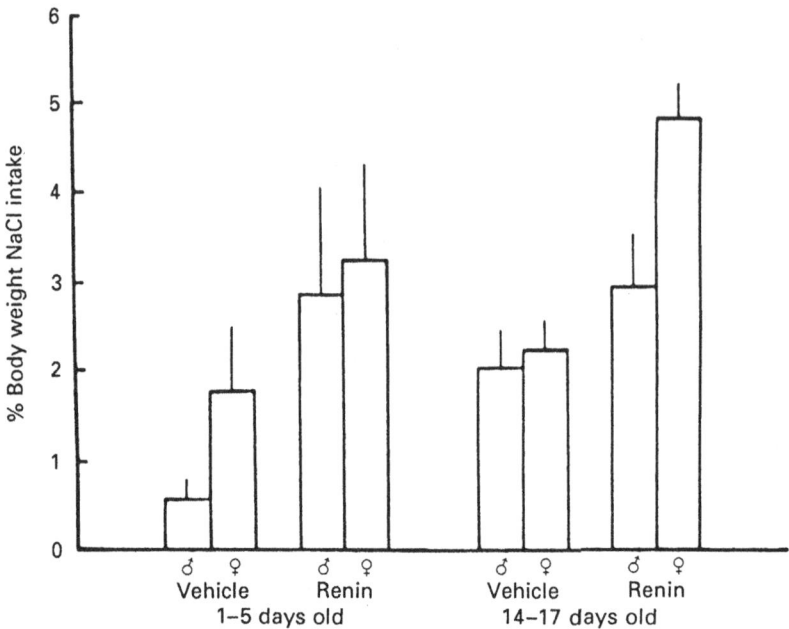

Fig. 2.2. Renin-induced NaCl intake in suckling pups 1–5 and 14–17 days old. The columns represent mean intakes expressed as a percentage of body weight (from Leshem & Epstein, 1989).

into the third ventricle, and then a short time thereafter are intraorally infused with sodium or other solutions. Their body weights are measured as an index of their ingestion; excretion is prevented as described in Chapter 1. Renin is clearly natriorexigenic early on in the neonate's life (e.g. Leshem & Epstein, 1989; Fig. 2.2) as is angiotensin itself (Thompson & Epstein, 1991). This mature ingestive response occurs before the actual expression of the rat's behavioral regulation of body fluids. That is, at this stage in development, the neonate is still suckling, and ingestion is the result of controls by the pup and its dam. Suckling, it should be noted, is not a prelude to adult ingestive behavior (e.g. Friedman *et al.*, 1981), although adult-like ingestive behavior can be expressed in the neonate. Thus, the demonstration of adult salt ingestion to natriorexigenic treatments, as I indicated in Chapter 1, is another example of an ingestive behavior that is precocious but is normally not expressed by the neonate.

Angiotensin-induced sodium hunger has been demonstrated in several other species. For example, the omnivorous pig not only increases its water but also its sodium intake when given renin or angiotensin into the third ventricle (Mutter *et al.*, 1984). Angiotensin induced sodium intake has also been demonstrated in mice (Denton *et al.*, 1990; Weisinger *et al.*, 1990). Moreover, the hunger for sodium is also expressed in a nonmammalian species. The pigeon, for example, also responds to either centrally administered renin or angiotensin by increasing its salt ingestion as well as its water intake (Epstein & Massi, 1987). The response is also dose dependent. Sheep also respond to centrally administered angiotensin (into the third ventricle) by ingesting sodium (see Denton's review, 1982). However, this ingestion of sodium may largely be due to the natriuretic effects of angiotensin (Chapter 4).

It is therefore clear that angiotensin administration results in increased sodium ingestion, which complements the increase in water intake. The hormone plays a dual behavioral function in the maintenance of body fluid homeostasis that complements its known physiological actions. The mechanisms of action will be discussed in Chapter 4.

MINERALOCORTICOID-INDUCED SODIUM HUNGER

Mineralocorticoid-induced sodium hunger is a paradigmatic example of a hormone-induced behavior. Its history began with Richter. He (1941) showed that adrenalectomized rats treated with deoxycorticosterone (DOCA), a mineralocorticoid, decreased the salt ingestion that resulted from the sodium loss (see Chapter 4). This is because, with DOCA acting on the kidney, rats do not lose sodium, and therefore cease to ingest it. What he was surprised to find was that the salt appetite returned again as the dosages of DOCA were increased, and that adrenally intact rats treated with DOCA ingested salt (e.g. Rice & Richter, 1943; Braun-Menendez, 1952). What these results suggest is that DOCA, a precursor of the mineralocorticoid ALDO, can increase salt ingestion when given alone. This is similar to giving angiotensin alone and generating a hunger for sodium; or giving estrogen or progesterone and eliciting lordosis behavior (Pfaff, 1980).

This area of inquiry lay dormant until the middle 1960s. The next figure, Fig. 2.3, depicts clearly the effects of DOCA (or DOCT) on adrenalectomized and adrenally intact rats from the work of Wolf(1965, see also Fregly & Waters, 1966). ALDO, the natural mineralocorticoid, also produces natriorexigenic effects in adrenally intact rats (Wolf, 1964b), and in adrenalectomized rats (Fregly & Waters, 1966, Fig. 2.4). Therefore, both ALDO and DOCA in the adrenalectomized rat reduce salt intake at therapeutic doses, which decreases the sodium output from the kidney, with higher doses, the appetite is once again reinstated. Importantly, while mineralocorticoid-treatment provokes a natriuresis or an 'escape' phenomenon (Chapter 4) over repeated days, this initial natriorexigenic effect is not related to sodium loss (Wolf, 1965). The appetite is independent of the sodium loss, though the natural conditions are that sodium loss results in the release of the mineralocorticoids which act to conserve sodium and then act on the brain to trigger salt ingestion (see Chapter 4). In other

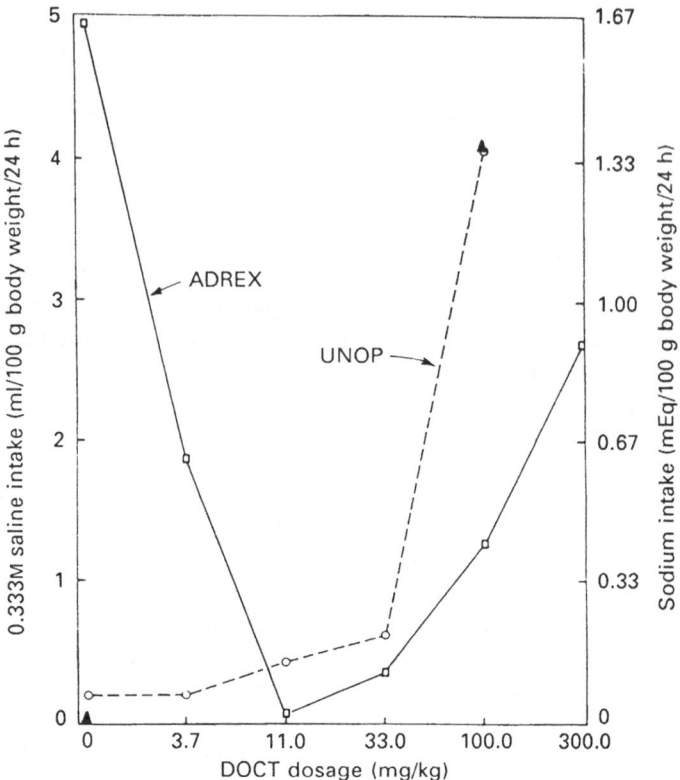

Fig. 2.3. Hypertonic (0.333M) saline intake of adrenalectomized, unoperated, and sham-operated (two solid triangles) rats as a function of DOCT dosage. Sodium intake is shown in terms of milliliters of solution consumed on left ordinate and in terms of milliequivalents of sodium consumed on right ordinate (from Wolf, 1965).

words, the dual function of the mineralocorticoids is clearly expressed: first to decrease sodium excretion and second to send signals to the brain that initiate salt ingestive responses.

Several hormonally induced behavioral responses were thought to be due to learning (e.g. insulin and feeding behavior; Woods *et al.*, 1977). And a dispute emerged as to whether the appetite for salt, that results from mineralocorticoid actions, is learned. Weisinger & Woods (1971) claimed that mineralocorticoid-induced salt appetite is learned, unlike sodium deficiency-induced salt appetite which is innate (Weisinger, 1975; Chapter 1). They hypothesized that, during the course of an animal's life, the animal would learn associations between body sodium deficiency and elevated aldosterone levels. But, this was later shown not to be the case. Rats raised on a high salt diet (but with normal body sodium, and

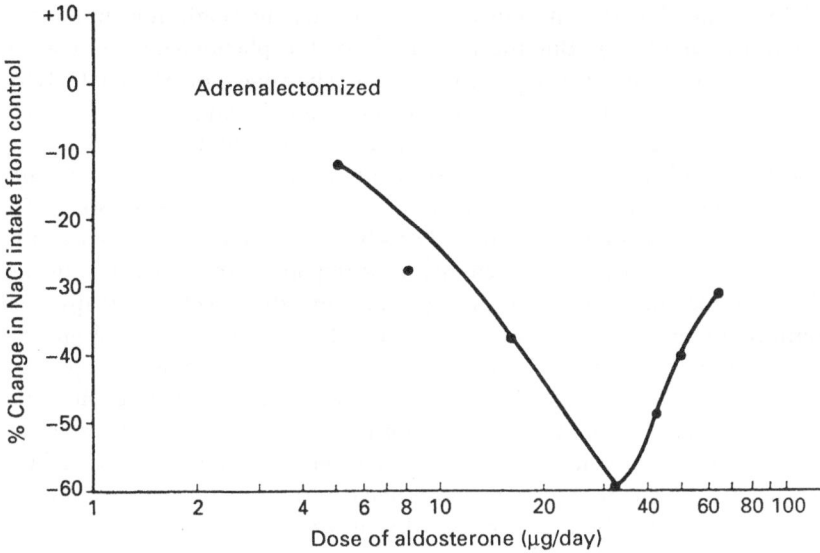

Fig. 2.4. Percentage change in intake of 0.15M NaCl solution of aldosterone-treated adrenalectomized rats from control adrenalectomized rats at each dose level of aldosterone administered (from Fregly & Waters, 1966).

which retained body sodium to the mineralocorticoid treatment) still manifested a salt appetite to the treatments (Schulkin, 1978). Therefore, mineralocorticoid-induced appetite, just like depletion-induced salt appetite, is innate (Chapter 1).

Mineralocorticoid receptors are known to appear in the brain by about 12 days of age (Coirini *et al.*, 1985; see Chapters 4 and 5 for a further discussion of mineralocorticoid receptors), and the hunger for sodium from systemic treatment with ALDO appears at this time (Thompson & Epstein, 1991). Moreover, recall from Chapter 1 that rats will demonstrate motivated behaviors for salt following mineralocortocoid treatment. They will labor to obtain salt (Wolf & Quartermain, 1966; Quartermain & Wolf, 1967), and will ingest salt adulterated with quinine. Mineralocorticoid-treated rats will also run down a runway for salt (Shulkin *et al.*, 1985*a*; Chapter 1). The running speed is dependent upon the level of mineralocorticoid treatment and the concentration of the salt. Mineralocorticoid-induced salt appetite is specific for salty tasting solutes (Chapter 3).

There are clear species differences to mineralocorticoid-induced salt hunger. Mineralocorticoid-induced sodium hunger was thought to be minimal in sheep (Denton & Nelson, 1970). But a later study demonstrated a sodium hunger with high doses of DOCA (Hamlin *et al.*, 1988). It is also difficult to get rabbits to exhibit mineralocorticoid-induced salt appetite, although not impossible (Denton et al., 1969). But the magnitude of the phenomenon is small in comparison with that demonstrated in the rat. The mineralocorticoid, DOCA, can increase the salt intake of hamsters and gerbils (Wong, 1977; Fitts et al., 1983), but the effect is difficult to demonstrate in the mouse (Rowland & Fregly, 1988c). The phenomenon is not just confined to mammals. Pigeons also demonstrate mineralocorticoid-induced salt intake (Epstein & Massi, 1987). Thus, the pigeon responds vigorously to both the mineralocorticoid and to angiotensin. It is clear that the animal most responsive to the natriorexigenic effects of the mineralocorticoids is the rat, and the effect is perhaps even confined to specific strains of rats (Rowland & Fregly, 1988a). While the appetite is less pronounced in other species, it is still somewhat present. Therefore, while the hormone is not necessary for the appetite, since the adrenalectomized rat does not have the hormone, one has some confidence that, under conditions in which the mineralocorticoids are elevated, such as sodium deficiency in the natural state, the hormone contributes in the mobilization of the search for, and the ingestion of, salty substances.

THE SYNERGY OF ANGIOTENSIN AND THE MINERALOCORTICOIDS

There are a number of examples of two or more hormones being combined and producing behavioral effects greater than the actions of either hormone acting alone (e.g. estradiol or progesterone on sexual receptivity, Pfaff, 1980). The hormones of sodium homeostasis, angiotensin and ALDO are paradigmatic of the principle of synergy. In fact, the natural occurrence is for both hormones to be elevated at the same time; that is, during sodium deprivation, both hormones are elevated to conserve sodium and to act on the brain to generate an appetite for salt.

Braun-Menendez (1952), experimenting with many combinations of hormones, discovered that the combination of renin and DOCA in rats produces a large salt appetite. This effect was additive, however, not synergistic. Braun-Menendez gave both hormones systemically. The salt ingestion amounted to adding the cumulative intake of either hormone given alone. When angiotensin is given systemically along with the mineralocorticoids, there is no synergistic effect on salt intake (Sakai & Epstein et al., 1990a).

But, when the peptide is administered centrally and the mineralocorticoid is administered systemically, one sees a clear synergistic effect on salt intake. Fluharty & Epstein (1983), in a beautiful set of experiments, showed that centrally administered angiotensin in concert with systemic DOCA produces a clear synergistic effect on salt ingestion (Fig. 2.5). Similar results have been found when adult rats are treated with angiotensin and ALDO instead of angiotensin and DOCA (Sakai, 1986; Goldman, Epstein & Schulkin, unpublished observations), or, in rat pups intraorally infused with NaCl

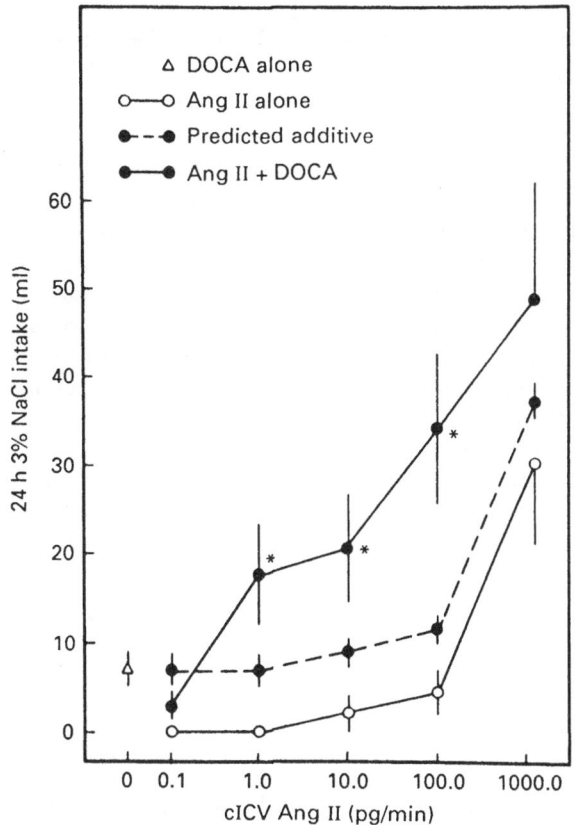

Fig. 2.5. Cumulative mean intake of 3% NaCl during 24 h cICV infusion of various doses of Ang II alone (*left panel*) or in animals additionally treated with DOCA (*right panel*). (Representative standard error bars are also shown. Ang II = angiotensin II; DOCA = deoxycorticosterone; cICV = continuous intracerebroventricular) (from Fluharty & Epstein, 1983).

(Thompson & Epstein, 1991). That is, rats pretreated with mineralocorticoids daily by subcutaneous injection, at doses which, by themselves, do not elevate salt intake, do increase their salt intake dramatically when they are injected intracerebrally with angiotensin. This dose of angiotensin, by itself, does not generate salt intake.

The synergy of cerebral angiotensin and the mineralocorticoids also generates motivated behavior. Recall the runway described in Chapter 1. We know that both mineralocorticoid, and for that matter, angiotensin, will generate running behavior for salty ingesta. It is not surprising then that the synergy treatment not only increases salt ingestion but also will motivate rats to run down a runway for the taste of salt (Zhang *et al.*, 1984).

But, again, there are species differences. The synergy effect has not been demonstrated in several other mammalian species. The sheep does not show synergistic effects to central angiotensin and the mineralocorticoids (Weisinger *et al.*, 1986). But then, as you will see in Chapter 4 and 5, sheep seem to use slightly different, though overlapping, mechanisms in the genesis of sodium hunger. The sheep is responsive to changes in brain sodium. The synergy effect has not been demonstrated in the dog, but then the dog is hardly responsive to any kind of natriorexigenic treatment. In fact, it is widely believed that the carnivorous dog, which gets its sodium needs from the prey that it kills, does not have a sodium hunger at all. It is not responsive to DOCA or to angiotensin (Ramsay & Reid, 1981). But as I noted in Chapter 1 it will, under sodium depletion produced by diuretics, increase its salt intake. Perhaps sodium sensors in the brain play a role in the modest sodium hunger in the dog (Chapters 4 and 5). In any event, the phenomenon of sodium hunger is expressed weakly in carnivores where there was not selective pressure for such behavioral mechanisms. But the synergy of angiotensin II and ALDO is not just confined to the rat. It has been demonstrated in the pigeon. That is, the pigeon seems very similar to the rat in the voracity of its response to the synergistic effects of angiotensin and aldosterone (Massi & Epstein, 1990).

Thus, the synergy phenomenon of cerebral angiotensin and systemic mineralocorticoid holds strongly in the rat and pigeon, but not for the sheep and dog. More species need to be tested.

ATRIAL NATRIURETIC FACTOR

As hormones can generate behavior, they can also inhibit or satiate behavior. Cholecystokin (CCK), for example, a gut peptide hormone, is thought to contribute towards the satiation of food hunger (Gibbs *et al.*, 1973). Atrial natriuretic factor (ANF), a peptide hormone from the heart, perhaps contributes towards the satiation of water and salt intake associated with extracellular or sodium excesses. As I indicated, ANF, found in both the atrium of the heart and in the brain, is often, but not always, an antagonist of the renin–angiotensin–aldosterone system physiologically and plays an important role in the excretion of excess sodium in extracellular fluid. These facts and others will be discussed at greater length in Chapters 4 and 5.

For now, consider that ANF modulates the behavioral expression of water and salt ingestion that results from angiotensin action. When injected into the cerebral ventricles, it reduces the ingestion of water that results from cerebral ventricular angiotensin injections. Carbachol-induced thirst is not affected by this treatment (Fitts *et al.*, 1985*a*). This may be an important fact, for it suggests that the effects of ANF are specific for the elicitation of drinking to angiotensin, which is known to be different from the mechanism that elicits carbachol-induced drinking. Perhaps the reduction in water intake is due to the

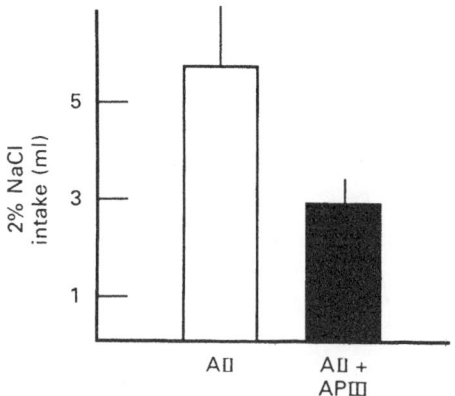

Fig. 2.6. Effects of atriopeptin III (APIII) on angiotensin II (AII)-induced NaCl intake. APIII (4 nmol) and AII (100 pmol) were given intraventricularly. Fluid intake was measured over a 1 h period. (from Masotto & Negro-Vilar, 1985).

interference of the angiotensinergic signal, and its role in extracellular fluid regulation.

Atrial natriuretic factor given directly to the cerebral ventricles also reduces the salt intake of sodium–depleted rats or rabbits, but not ANF to the systematic circulation (Fitts *et al.*, 1985*a*; Tarjan *et al.*, 1988). Importantly, ANF administered to the third ventricle reduces the water and salt ingestion that occurs following the delivery of angiotensin direct to the third ventricle (Antunes-Rodriquez *et al.*, 1986; Masotto & Negro-Vilar, 1985; Fig. 2.6). These observations about ANF suggest that it is central ANF that produces the antinatriorexigenic action during central angiotensin administration. In contrast to the above results, the salt appetite that results from mineralocorticoid treatment is not as affected by cerebrally injected ANF. In addition, the enhanced salt appetite of female rats, subject to a past history of sodium depletion treatments described in Chapter 1, is also not affected by ANF injections into the third ventricle (Massi & Schulkin, unpublished observations).

These results, when taken together, suggest that ANF, when injected into the brain, but not when injected systemically, reduces the salt intake due to renin–angiotensin.

Summary
The three hormones that regulate extracellular fluid sodium balance are critically involved in the behavioral regulation of salt ingestion. Angiotensin and ALDO figure prominently in the genesis of the appetite, whereas ANF is important in the inhibition of the angiotensinergic contribution to the appetite (see also Chapters 4 and 5).

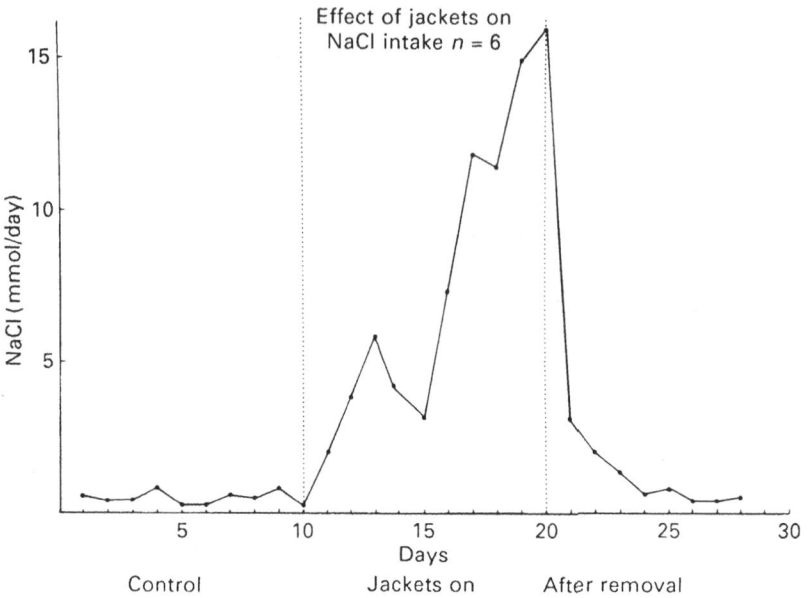

Fig. 2.7. Mean daily intake of 500 mmol/1 sodium chloride solution by wild rabbits before, during, and after attachment of jackets to their backs (from Denton *et al.*, 1984).

PITUITARY–ADRENAL CONTRIBUTION: ACTH AND THE GLUCOCORTICOIDS

Hans Selye, in his classic description of the adaptation to stress, pointed to the important role of the pituitary adrenal axis (1952). The pituitary gland, the master gland of the endocrine system, is activated under a number of conditions. One behavioral result of pituitary activation is salt ingestion. This is particularly true under conditions of duress. The activation of the pituitary gland affects the adrenal gland (see Chapter 4), resulting in the release of the mineralocorticoid and glucocorticoid hormones. Specifically, ACTH from the pituitary gland results in these steroids being synthesized and released from the adrenal gland, with behavioral consequences.

In fact, during stressful or laborious conditions animals (e.g. rabbits) are known to increase their salt ingestion. They do so when housed in small surroundings with other animals, or when movement is prohibited (Denton *et al.*, 1984a, Fig. 2.7). During each of these conditions, the pituitary–adrenal axis is activated.

Interestingly, ACTH, when given by itself, subcutaneously generates an appetite for salt in several species – sheep, rabbits, mice and rats (e.g. Weisinger

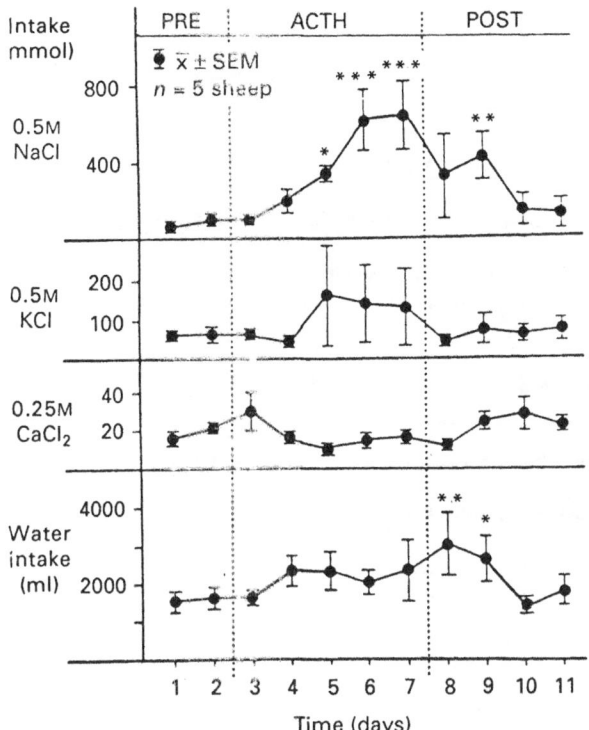

Fig. 2.8. Mean intakes of salt following ACTH injections (90 i.u./day) (after Weisinger *et al.*, 1980).

et al., 1978; Denton, 1982; Fig. 2.8). The effect in rats and sheep, however, requires an intact adrenal gland. This is not the case for the rabbit. That is, adrenalectomized rats and sheep maintained with therapeutic levels of mineralocorticoid and glucocorticoid hormones do not ingest salt after ACTH administration. This suggests that ALDO, perhaps with the participation of corticosterone, the major glucocorticoid hormone, is significantly involved in ACTH-induced sodium hunger in the rat and sheep. By contrast, rabbits whose adrenal glands are removed, and who are maintained with therapeutic doses of ALDO so that they do not lose body sodium, increase their salt intake after ACTH administration. Therefore, the ACTH-induced sodium hunger in this species is independent of the adrenal steroid hormones.

Glucocorticoids, when given alone, can also generate an appetite for salt in several species (see Denton, 1982; rabbits and sheep). The results from one such study are shown in Fig. 2.9. The effects are even greater with ACTH. But, there are species differences with regard to whether the glucocorticoid will elicit salt

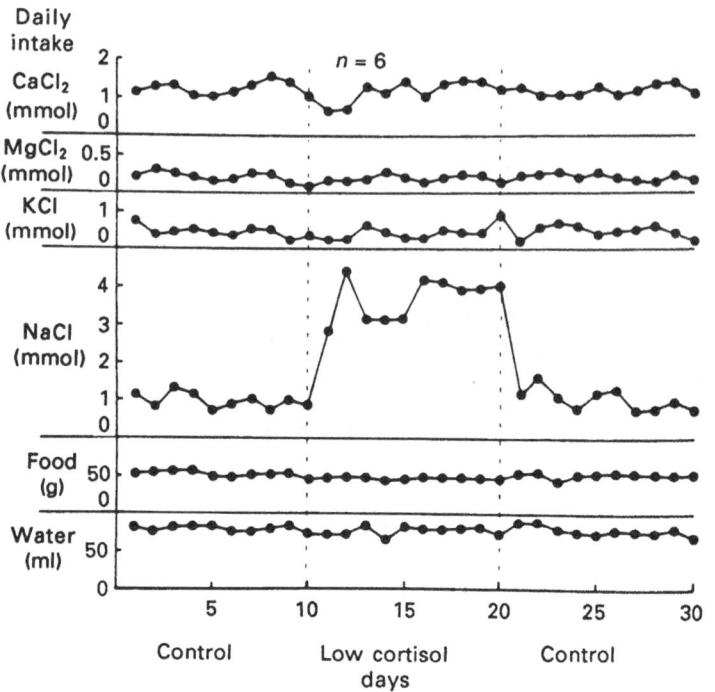

Fig. 2.9. Effect on adrenalectomized wild rabbits of reduction of cortisol administration from the maintenance dose (0.75 mg/day) to 0.375 mg/day on the mean daily intakes of food, water, 500 mM sodium and potassium chlorides, and 250 mM magnesium and calcium chlorides. Maintenance treatment with DOCA (0.1 mg/day) was continued throughout (from Denton, 1982).

intake. While the glucocorticoids do not elicit a sodium hunger in rats when given alone (e.g. Wolf, 1965) they do so when given with the mineralocorticoids. In an interesting experiment, Wolf (1965) showed that, in adrenalectomized rats treated with different doses of DOCA (restoring sodium balance), high doses of DOCA combined with corticosterone resulted in even greater salt intake (Fig. 2.10). In a similar study (Coirini et al. 1988), dexamethazone treatment increases substantially the salt intake of DOCA-treated rat, or of ALDO-treated rats. One possible reason for the glucocorticoid contribution in the ingestion of salt, elaborated in Chapters 4 and 5, is that the glucocorticoids increase the level of mineralocorticoid receptor sites in the forebrain regions of the brain, and therefore increase the natriorexigenic actions of the mineralocorticoids.

These results when taken suggest that stress-induced sodium hunger may be related to a heightened pituitary–adrenal secretion. This natriorexigenic effect

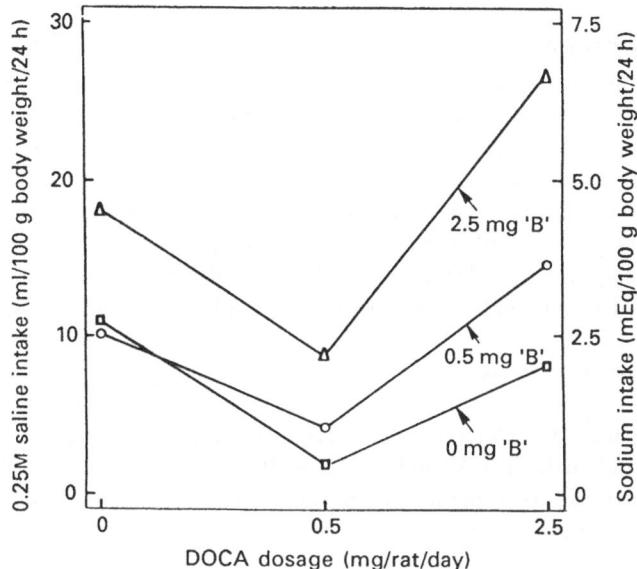

Fig. 2.10. Saline (0.25 M) intake of adrenalectomized rats injected daily with 0, 0.5, 2.5 mg of corticosterone acetate (B) as a function of daily DOCA dosage. Sodium intake is shown in terms of milliliters of solution consumed on left ordinate and in terms of milliequivalents of sodium consumed on right ordinate (From Wolf, 1965).

should be distinguished both conceptually and empirically from the natriorexigenic effects of the hormones of sodium homeostasis – angiotensin and ALDO. Thus, in summary, ACTH and the glucocorticoid hormones contribute to the genesis of sodium hunger that is probably related to stress. The glucocorticoid hormones also contribute to the natriorexigenic effects of the mineralocorticoid hormones.

SEX DIFFERENCES IN SALT INGESTION UNDER HORMONAL CONTROL

Clearly, hormones have profound effects upon salt ingestion. In this section, I will review the dramatic differences in the salt (and water) ingestion of male and female rats. This phenomenon is dependent upon the hormonal milieu during critical stages in the development of gender (organizational changes in the brain and on behavior) and upon the activation of behavior during the course of the animal's adult life. The warranted hypothesis is that females are under great danger of losing critical body fluids during reproduction and therefore have evolved special mechanisms for the maintenance of body fluids. Salt intake is a sexually dimorphic behavior.

FLUID INGESTION

There have been a number of observations showing that female rats respond differently than male rats in the ingestion of sweet solutions (Wade & Zucker, 1969a). The response appears to be hormonally mediated. Female rats ingest greater amounts of sweet solutions, a response altered by the gonadal steroid hormones; ovarectomy during the neonatal period of development (before the 12th day of age) abolishes this phenomenon. In addition, female rats treated with androgens during this critical stage of development also do not show this elevated ingestion of sweet solutions when mature (Wade & Zucker, 1969b; Zucker, 1969).

Female rats are particularly responsive to extracellular dipsogenic signals. They ingest greater amounts of water than male rats in response to angiotensin II injections (Kaufman, 1980). Depletion of extracellular fluid also elicits greater drinking responses in females (e.g. Vijande et al., 1977). After all, it is the loss of extracellular fluid that occurs in the female during the demands of reproduction, so she is more sensitive to dipsogenic challenges that alter extracellular fluid balance.

There is further evidence that the sensitivity of the female rat to changes in extracellular fluid balance is specific. Increased drinking to dipsogenic signals that are not related to the maintenance of extracellular fluid (e.g. carbachol and intracellular dehydration) do not have different effects in male and female rats (Jonklass & Buggy, 1984, 1985). Moreover, the exaggerated responsiveness to extracellular stimuli appears to be hormonally mediated, since neonatal androgenization abolishes this effect (Kucharczyk, 1984a, b).

It is during the lactation phase of the reproductive event that there is great body fluid loss in the female, and it is therefore quite interesting that one of the hormones of lactation, prolactin, is also known to elicit drinking behavior (Kaufman et al., 1981). When prolactin is combined with extracellular depletion, the female rat's drinking response is much greater than that of the males (e.g. Kaufman and Mackay, 1983). In a moment, you will see that prolactin contributes to the genesis of salt ingestion tied to the regulation of sodium and extracellular fluid balance during lactation.

But, water intake is also known to decrease during the female's biological cycles. That is, water intake (like food intake) is also known to decrease during the estrous phase of the cycle (e.g. Jonklass & Buggy, 1984, 1985). Estrogen contributes to this decrease in ingestive response. Estrogen treatment decreases angiotensin-induced water drinking (Fregly, 1980; Findlay et al., 1979). Ovarectomy abolishes this reduction in ingestion, and estrogen treatment once again reinstates it.

Interestingly, the reduction of water intake during the estrous phase is related to the actions of angiotensin. Angiotensin-induced drinking is reduced, whereas intracellular dehydration thirst is not affected by changes in the estrous cycle. Carbachol-elicited thirst behavior is also not affected by the estrous cycle (Jonklass & Buggy, 1984; Kucharczyk, 1984a, b).

As shown earlier, angiotensin, the hormone of extracellular fluid balance (Fitzsimons, 1979) recall elicits both water and sodium ingestion, suggesting again that the drinking behavior that is most affected by the estrous cycle is angiotensin-induced drinking related to extracellular fluid balance.

There is more evidence for the claim that estrous specifically affects the angiotensinergic signal in the maintenance of extracellular fluid balance. Central angiotensin induced-thirst is also affected by central estrogen treatment; interestingly, the treatment has no effect on intracellular dehydration thirst in female rats (Jonklass & Buggy, 1985). In addition, estrogen treatment has no effect on thirst elicited by either carbachol or intracellular dehydration in male rats. These results further suggest that the effects of estrogens are specific to extracellular fluid regulation and perhaps to the angiotensinergic contribution.

It is also known that injections of angiotensin directly into the preoptic region result in reduced drinking during the estrous cycle, while injections elsewhere in the brain do not (Jonklass & Buggy, 1985). This latter result suggests that areas within the preoptic region may be importantly involved in the angiotensin-related reduction of water intake during the estrous cycle.

In fact, there is direct evidence for this claim. Female rats treated with estrogen in the medial preoptic region showed decreased responsiveness to centrally administered angiotensin. Estrogen implanted elsewhere in the brain, e.g. in the ventral medial hypothalamus, did not result in this reduction in angiotensin-induced drinking (Jonklass & Buggy, 1985). The treatment had no effect on males. In addition, neonatal androgenization abolished this effect in the females. Moreover, it was also demonstrated that systemic estrogen treatment reduced angiotensin receptors in a block of tissue that included the preoptic region, but not elsewhere in the brain. These results, when taken together, suggest that the preoptic region, which includes a sexually dimorphic nucleus (see Chapter 5), may be at the basis for this differential response to angiotensin-induced drinking during the estrous cycle and the response to extracellular fluid maintenance, in general, during the course of the female's daily life (Chapter 5). In addition, neonatal androgenization results in a greater volume of the sexually dimorphic preoptic nucleus in addition to the medial bed nucleus of the stria terminalis and medial nucleus of the amygdala (Chapter 5). As the reader will see, these sexually dimorphic brain regions may play a role in the female's ingestion of salt and in mineralocorticoid-induced sodium hunger. In chapter 5, I will suggest that the evolution of mineralocorticoid-induced sodium hunger is tied to the demands of reproduction, and therefore, perhaps, females share a common steroid-sensitive circuit in the forebrain.

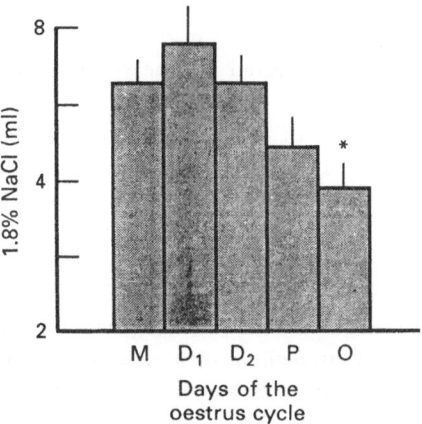

Fig. 2.11. Variation in 1.8% saline ingested across days of the estrous cycle. M = metestrus, D_1 = diestrus 1, D_2 = diestrus 2, P = pro-estrus, O = estrus. (from Danielsen & Buggy, 1980).

SALT INTAKE

The estrous cycle also results in the reduction of salt ingestion. After all, the complement of the reduction of water intake is the reduction of sodium intake. For example, in sheep, salt intake declines during the estrous cycle (Michell, 1975), as it does for rats (Danielsen & Buggy, 1980). And angiotensin-induced salt intake in rats, like water intake in rats, also decreases at estrous (Fig. 2.11). This, then, is the full complement of the effects of estrous on water and sodium intake.

PREGNANCY AND THE HORMONES OF SEXUAL REPRODUCTION

A number of animals are known to search for, and ingest, salt licks during the reproductive season. As indicated in Chapter 1, it is the female during the reproductive season that is often seen at mineral licks, since, during this time, there are greater demands on the female for minerals.

In the laboratory, when not in the estrous cycle, female rats typically ingest more salt than male rats. This is particularly true during pregnancy. Richter charted the ingestive course of female rats and noted their ingestion of various minerals. Figure 2.12 shows that their ingestion of sodium began to increase shortly after they mated (Richter & Barelare, 1938). Other minerals were also ingested in greater amounts. Perhaps what one sees in this context is the onset of a mineral hunger, of which sodium ingestion is just one expression.

During pregnancy, female rats, and other animals, are known to lick their genitals to reabsorb lost sodium (Steinberg & Bindra, 1962). The genital licking

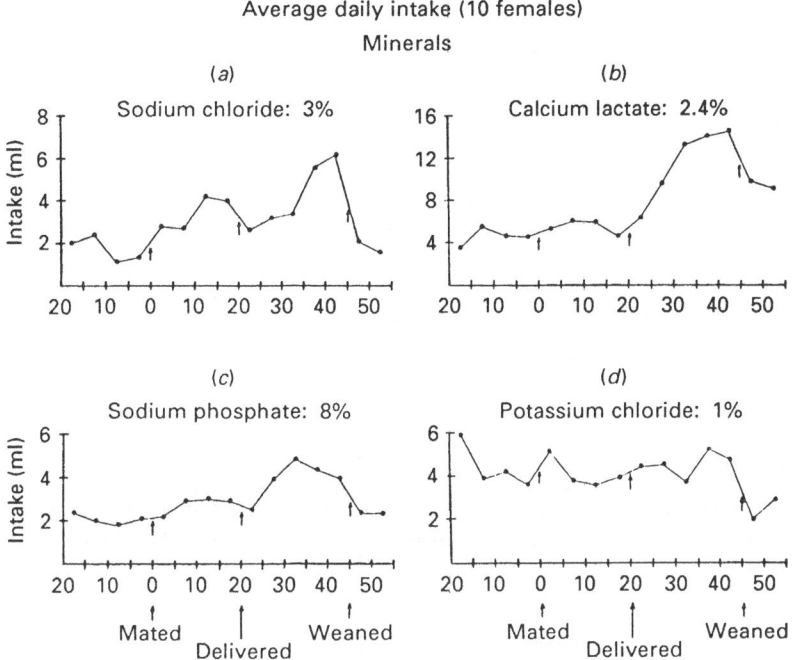

Fig. 2.12. Curves showing average daily intake of the various mineral solutions in 5-day periods for 20 days before mating, for 20 days of pregnancy, for 25 days of lactation, and for 10 days after weaning (from Richter & Barelare, 1938).

declines again after the pups are weaned. Interestingly, during pregnancy and lactation, ALDO is elevated in the body (Quirk *et al.*, 1983), perhaps contributing to the physiological redistribution and conservation of sodium as well as to the elevated salt intake.

This salt ingestion phenomenon has also been analyzed in wild rabbits. A dramatic increase in sodium (and other minerals) intake occurs during pregnancy and is at its zenith during lactation (Denton & Nelson, 1971). The salt intake is related to the size of the litter and the demand on the mother, which really comes down to the demand for sodium. Figure 2.13 depicts the relationship between the litter size and ingestion of salt. The greater the litter, the greater the ingestion.

As I indicated earlier, there are studies in the rabbit that indicate that hormonal factors that influence lactation can elicit an appetite for salt (see Denton, 1982). Wild rabbits, that are treated systemically with prolactin, increase their salt intake (Fig. 2.14). Oxytocin is also natriorexigenic in the rabbit (Fig. 2.15). In both cases, the salt intake is independent of sodium loss.

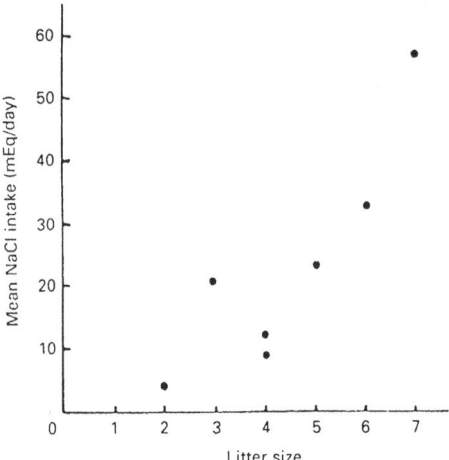

Fig. 2.13. Relationship between number of babies in individual litters and the mean daily NaCl intake of the respective mothers during lactation (from Denton & Nelson, 1971).

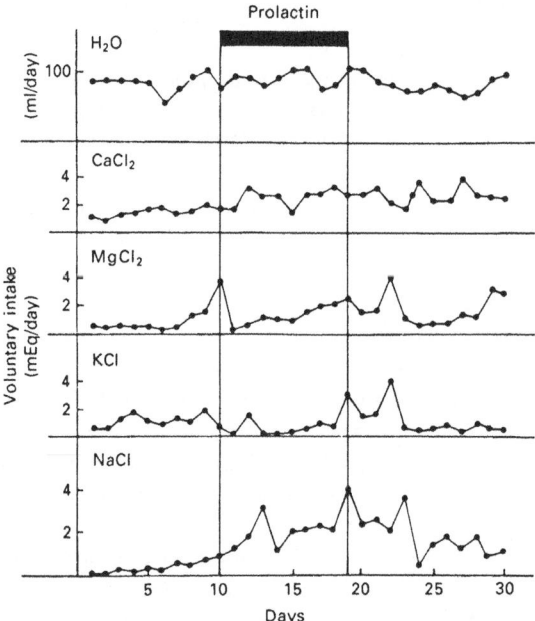

Fig. 2.14. Effect of 50 i.u. prolactin per day on the mean voluntary intakes of electrolyte solutions and water (from Shulkes *et al.*, 1972).

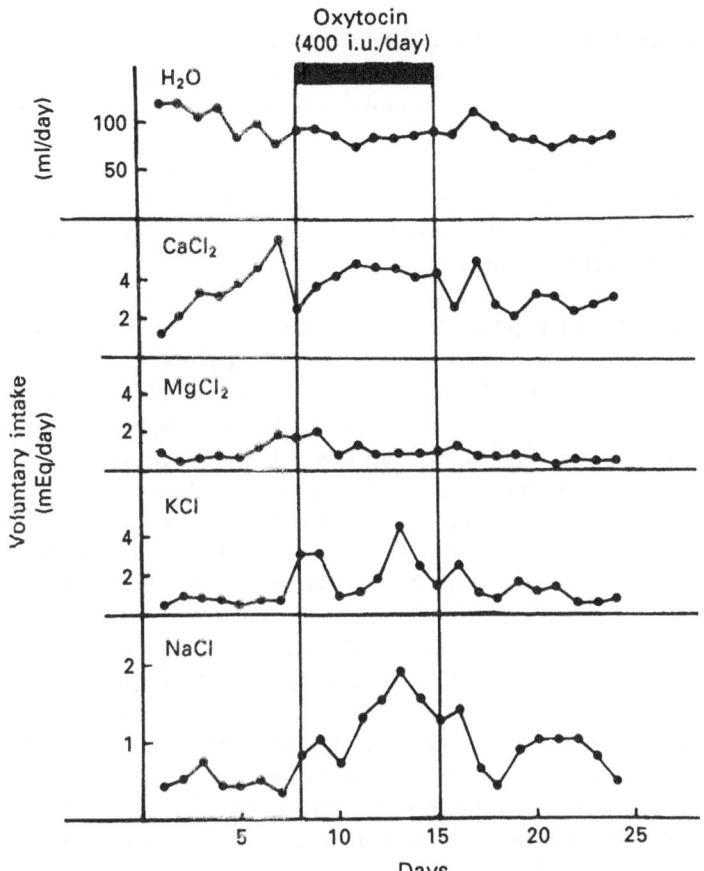

Fig. 2.15. Effect of 400 mμ oxytocin per day on the mean voluntary intakes of electrolyte solutions and water (from Shulkes *et al.*, 1972).

Therefore, it seems fairly clear that the hormones involved in lactation are able to elicit an appetite for salt.

The phenomenon of increased salt intake during pregnancy and lactation is also found in another mammal in addition to the rat and the rabbit: the mouse (McBurnie *et al.*, 1988). Additional studies are required to see if this enhanced salt intake during the reproductive cycle occurs in birds. Recall that the pigeon is responsive to natriorexigenic treatments. Perhaps even carnivores express an enhanced craving for minerals during the reproductive cycle. There is some evidence in pregnant humans of enhanced craving for sodium and other minerals (Brown & Toma, 1986).

CONTRACEPTIVES INFLUENCE SALT INTAKE

Additional factors that influence the female's hormonal milieu affect salt ingestion. Long-term exposure of rats to oral contraceptives increases salt intake (Fregly, 1973). Rats treated with envoid, an oral contraceptive, in their food for up to one month and then presented with a salt solution, ingest it avidly. This effect is seen in both male and female rats, and in ovarectomized female rats. The longer the time of exposure to the envoid, the greater the salt intake. Minerals other than sodium were also ingested.

HORMONAL INFLUENCES IN GENDER DEVELOPMENT AFFECT SALT INGESTION

As in sexual preference, the hormonal mileu during a critical stage of development profoundly influences salt ingestion. Female rats, and monkeys also typically ingest more salt than male rats even when they are not under the influence of the ovarian hormones or during states of pregnancy or lactation (Krecek et al., 1972; Krecek, 1973; Fig. 2.16). In addition, there is some evidence that female Rhesus monkeys ingest greater amounts of sodium than male rhesus monkeys (Schulkin et al., 1984; Fig. 2.17). It should be further exaggerated during the reproductive cycle.

Testosterone seems to be involved in this gender-related difference in the ingestion of salt. Female rats treated with testosterone at 2 days of age ingested salt as if they were males when tested in adulthood (Krecek et al., 1975). After 12 days of age, the treatment had no effect. In addition, testosterone inhibits salt intake in adult rats (Sakai et al., 1988). Male rats treated with estradiol at 12 days of age still had body weights lower than the control male rats, and when they were mature, did not behave towards salt as mature female rats do (Krecek et al., 1975). Castration in males before the critical stage in development (before 12 days of age) did not produce female-like inhibition of drinking to angiotensin following estrogen treatment (Jonklaas & Buggy, 1985), but it did produce female-like ingestion of salt in the male rat (Sakai et al., 1988). Therefore, some aspects of ingestive behavior are altered by the endocrine manipulations, while others are not.

Finally, the pineal gland, known for its involvement in the organization of biological rhythms in some species (see Rosenwasser & Adler, 1986), may play a role in the inhibition of salt intake. Pinealectomy can increase salt intake (Bliss & Bates, 1972). In fact, removal of the pineal gland, during the neonatal period, potentiates the female's avidity for salt; pinealectomy during adulthood is without effect (Krecek et al., 1975). In addition, if male rats are gonadectomized during the neonatal period, the salt intake that results from the removal of the pineal gland is potentiated (Zicha et al., 1972).

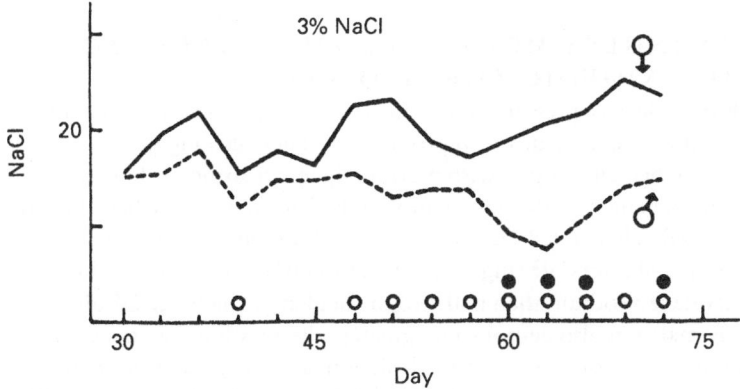

Fig. 2.16. Effect of age and sex on the relative intake of 3% NaCl solution. The fluid consumption is expressed in ml/100 g body weight per 3 days (from Krecek, 1973).

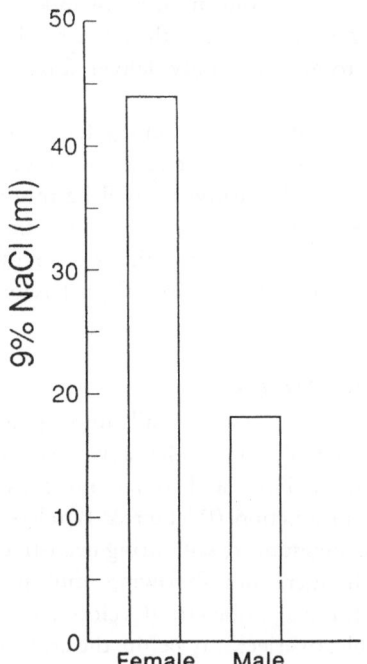

Fig. 2.17. Mature female and male rhesus monkey's salt intake (from Schulkin, Liebman, Ehrman, Norton & Ternes, 1984).

FEMALES INGEST MORE SALT THAN MALES TO NATRIOREXIGENIC TREATMENTS

There is clearly selective pressure on females to express an enhanced appetite for salt that is tied to the demands of reproduction. They also ingest more NaCl than male rats when challenged with natriorexigenic treatments. For example, virgin mature female rats also ingest more salt than male rats when they are sodium depleted. That is, female rats that are sodium depleted and then given access to sodium 24 hours later ingest greater amounts than the male, while they actually excrete less sodium than males to the depletion challenge (Wolf, 1982; Fig. 2.18). Female rats also demonstrate greater salt ingestion than males on the second sodium depletion over the first depletion, and the greater need-free salt intake that results from the sodium hunger (Sakai *et al.*, 1987; Chapter 1). The phenomenon is also manifested in the male that was gonadectomized during the critical stage of development (Sakai *et al.*, 1988).

This greater salt ingestion is not just confined to sodium depletion treatments, nor just to the rat. Thus, female hamsters and rats are also more responsive than male hamsters or rats to mineralocorticoid-induced salt appetite (Fitts *et al.*, 1983; Marini *et al.*, 1986, Chapter 5), and, even in the neonate, the female ingests more salt than the male does to intracerebrally delivered renin (see Fig. 2.2; Leshem & Epstein, 1989).

Thus, this enhanced avidity for, and greater intake of, salt in the female is demonstrated following natriorexigenic treatments. Further research should extend the range of animals tested. Perhaps female carnivores will be more responsive than males to natriorexigenic treatments, and will show that the evolution of sodium hunger is tied to the female and perhaps to the creation of steroid circuits in the forebrain including those that may suppress the behavior in males.

SUMMARY OF SEXUAL DIFFERENCES

Like other hormonally induced behaviors, salt ingestion is a sexually dimorphic behavior. Female rats ingest more salt than male rats under a variety of conditions. They ingest large amounts of sodium and other minerals, particularly calcium, under conditions of reproduction (Richter & Barelare, 1938; Woodside & Millelire, 1987). They also ingest more salt during need-free conditions, under conditions of sodium hunger, and following multiple treatments that induce sodium hunger. Female rats also obey the developmental laws of behavioral endocrinology; for example, ovariectomy before the critical stage in development reverses the elevated salt intake typically seen in virgin females. These behavioral results are therefore consistent with other behaviors that are organized by neuroendocrine function (Goy & McEwen, 1977). Thus, the results discussed in this section suggest that selective pressure on females

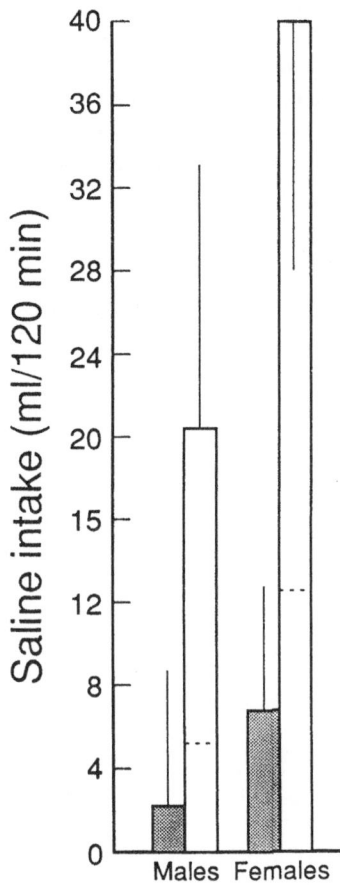

Fig. 2.18. Intakes of saline solution under need-free (baseline) condition (shaded bars) and under natriorexigenic condition (unshaded bars) for groups of male and female rats. (the bars represent mean 2-h intakes, with standard deviations). The dotted horizontal lines inside the unshaded bars represent mean amounts ingested during the first 10 min of testing (from Wolf, 1982).

through the course of evolution provided them with special hormonal mechanisms to insure stable sodium balance during pregnancy and lactation when there are greater demands for sodium and extracellular fluid balance.

GENERAL SUMMARY
This chapter has documented hormonal factors in the regulation of sodium ingestion, fluid intake, and other minerals. There are three broad classes of

events. The first is the defense of sodium and extracellular fluid balance. Three hormones are principally involved: angiotensin, ALDO and ANF. Second, in response to high arousal, the pituitary adrenal axis is activated; ACTH and the glucocorticoids elevate salt ingestion. The third set of hormonal factors are the hormones of sexual reproduction, which, in the female, are tied to the metabolic demands of pregnancy and lactation. Perhaps the evolution of sodium hunger principally evolved in this third context.

3 Gustatory contribution to salt intake

INTRODUCTORY

The search, identification, and ingestion of salt is perhaps the most studied example of how gustation organizes behavior. Animals use gustation, as a primary sense, in the exploration of their world. At certain times, they search for salty tastes over other gustatory enticements, e.g. sweets. Tasting is an active process (Halpern, 1983), and like other sensory events is under centrifugal control; gustatory nerves and lower gustatory regions are controlled, or informed, by more rostral brain sites. These sites orchestrate the behavioral search for, and acceptance of, salty tasting substances.

Gustation is an interface between the outside world and the internal milieu (Rozin & Schulkin, 1990), and is tied to homeostatic regulation. Gustation is the first event of the digestive process, and its importance in the identification of ingesta in the regulation of the internal milieu is uncontested.

Animals recognize taste properties rapidly. Rats, for example, can recognize a salty taste within milliseconds (Halpern & Tapper, 1971). Recall Chapter 1 where evidence was shown that rats can recall where salt was located and how it was perhaps obtained with just five small licks of NaCl.

The ingestion of salty substances, like other taste qualities, appears to be controlled through two independent palatability dimensions: one for acceptance and one for rejection of the salt (Berridge & Grill, 1983). This acceptance and rejection system is manifested at various levels of the neural axis, from the oral cavity to the cortex.

A salty taste serves as a marker or signal for mineral deposits. This is why, under laboratory conditions, one finds NaCl ingested by potassium-, calcium-, and zinc–deficient rats (Schulkin, 1986). In Nature, the salty taste signals that mineral requirements can be satisfied at a particular location. These are usually what we call 'salt licks'. The salty taste of sodium is linked to a specific receptor system. This innate hunger may organize other mineral hungers to search for salty tastes associated with mineral licks.

Picture the following: the sodium hungry animal searches for something salty. Oral gustatory changes in the detection and identification of saltiness prompt its search; and there are motivational and affective changes that potentiate the behavioral search. Gustatory and cephalic, in addition to postabsorptive factors, figure in the regulation of sodium, including the need-free preference for NaCl. The chapter begins with a discussion of the evolution of salt taste receptors, the detection and identification of saltiness, peripheral and central systems involved in its perception, salt taste development, changes in salt hedonics, preferences for sodium, species differences for salt taste, gustatory and cephalic factors in the satiation of sodium hunger, and, finally, the search for a salty taste. I emphasize behavioral analyses of sodium hunger throughout, for example, to clarify neurobiological issues.

EVOLUTION OF SALT TASTE RECEPTORS

Chemoreception is an ancient capacity that evolved over 500 million years ago. When animals moved from the sea onto the land, one of their problems was to find sources of sodium. The body cannot make sodium; animals must find and ingest it. When in the sea, the challenge was to excrete sodium. For land-dwellers, the issue is to conserve body sodium and to identify sodium in the environment.

One adaptation was the evolution of a gustatory system able to identify salty tastes. This adaptation is ubiquitous in terrestrial animals; it is found in invertebrates, amphibians, reptiles, birds, and, of course, mammals. Specific sodium receptors are found, for example, in frogs (Yoshii et al., 1986), and in blowflies (Dethier, 1968), in addition to in some mammals and birds. Sodium-specific gustatory receptors appeared early on in evolution, and actually preceded their actual use in the regulation of salt ingestion by behavioral means since invertebrates (blowflies, Dethier, unpublished observations) and reptiles (turtles, see Spigel et al., 1967) do not express a sodium hunger when depleted of sodium.

DETECTION AND INDENTIFICATION OF SALTY TASTES

When sodium hungry, it is obviously essential to detect or identify salt. This is the primary role of the gustatory system in sodium hunger. I begin by noting an important distinction in psychophysics.

There is a distinction in taste psychophysics between the detection of a gustatory stimulus, and its identification as having a particular taste quality. For example, Richter found that humans could detect NaCl at concentrations below that for which they eventually recognized them as salty (Richter & MacLean, 1939). The same holds for other taste qualities (Bartoshuk, 1979).

One question that emerges is the following: is the detection or recognition threshold for NaCl enhanced in the sodium-depleted animal? Richter seemed to think that it was. He found that adrenalectomized rats could respond to low concentrations of NaCl that rats with intact adrenals could not (Richter, 1939). To demonstrate this idea, he gave rats access to different concentrations of NaCl, or salty water, along with access to water. The experimental paradigm was a 24-hour intake test, and the NaCl solutions were given in ascending order of concentrations. Adrenalectomized rats started to ingest the salt at concentrations that were an order of magnitude lower than those for the intact rats. Richter's interpretation of this was that the sodium-hungry rat could detect or recognize the NaCl at 0.003%; the intact rat could detect or recognize it at 0.055%. Richter thought that changes in the sodium concentration which perfused the oral cavity accounted for this.

But, was it the case that the sodium hungry rat was simply responding to the salt taste because it now needed the sodium and not because its detection or recognition thresholds were altered? The issue is difficult to resolve. The intact and adrenalectomized rats that manifest a preference for 0.03 to 0.09% NaCl (Richter, 1939; Bare, 1949) can actually detect some taste quality at much lower concentrations: 0.003 to 0.009% NaCl (Carr, 1952), or perhaps even lower than that (Harriman & MacLeod, 1953). Sodium-depleted goats and calves ingest sodium salts at lower concentrations than nondeprived goats and calves (Bell & Williams, 1960). The same holds for mineralocorticoid-treated rats (Herxheimer & Woodbury, 1960); importantly, no such effects were found in the ingestion of sweet solutions. Moreover, acute water loading coupled with sodium loss in humans resulted in similar ingestive responses to sodium salts. Importantly, experiments looking at shock-motivated discrimination of salt taste suggest that thresholds are not different in the sodium-hungry rat (Carr, 1952), and that adrenal intact rats can, in fact, learn to avoid shock by detecting a taste property, again perhaps the saltiness, as low as 0.002% (Koh & Teitelbaum, 1961). These experiments highlight the importance of interactions between detecting or recognizing a salty taste and being motivated to drink NaCl.

The sodium that bathes the oral cavity from saliva is known to influence salt perception (e.g. Contreras & Catalanotto, 1980; Thrasher & Fregly, 1980). In fact, psychophysical experiments demonstrate that the concentration of sodium in the oral cavity affects the perception of salt in humans. When distilled water is applied to the oral cavity, the threshold for salt recognition is about 0.00014M. When it is not, it is about 0.0043M (McBurney & Pfaffmann, 1963). This probably reflects the different signal-to-noise ratios seen in the rinsed and unrinsed controls. In addition, the taste quality and intensity of NaCl changes, for example, depending upon the adaptation of the oral cavity to various concentrations of sodium (see Bartoshuk, 1974). NaCl can even taste sweet (Bartoshuk *et al.*, 1978). Thus, the higher the concentration of sodium in the oral cavity, the less sensitive one is to saltiness, and therefore there is an increased detection threshold. This is consistent with the fact that sodium deprivation in humans (Bertino *et al.*, 1981; Chapter 1) or acute sodium loss (McCance, 1936, 1938; Schmidt-Nielsen, 1964) perhaps actually decreases the sensitivity to salty tasting commodities (cf. Yensen, 1958, 1959 with Henkin *et al.*, 1963). However, it can also make the salty taste more pleasant (Beauchamp *et al.*, 1983; see also hedonics section of this chapter).

Sodium hunger can either decrease or increase the concentration of sodium that bathes the oral cavity (Denton, 1982). Yet, in both cases, one sees salt appetite. In addition, desalivated rats manifest a salt appetite both when sodium depleted and when mineralocorticoid treated (Vance, 1965; Wong & Kraintz, 1977). They also have an exaggerated saline preference (e.g. Vance, 1965; Wong & Kraintz, 1977). Moreover, detection and ingestion of sodium at low concentrations is achieved by the sodium–depleted desalivated rat (Wong & Kraintz, 1977). Thus, it must be something other than salivary sodium that is determining the ingestion, though, no doubt, it contributes. Moreover, it is also not a matter of sodium content that determines the initial response of the sodium deficient rat but the salty taste, since it will ingest equal amounts of 0.3M NaCl or 0.03M Na_2CO_3. These two salts are thought to taste similar (Morrison & Young, 1972), though the sodium content is very different. However, they are ingested in equal amounts; that is, sodium–depleted rats, offered a choice between two equally preferred and salty-tasting sodium salts, ingested equal amounts of the sodium salts, even though one sodium salt (0.3M NaCl) is five times richer in sodium than the other sodium salt (Morrison & Young, 1972). There appears to be something about the taste of salt that determines the ingestion (Morrison & Young 1972); I submit it is the saltiness.

The expression of a salt hunger, like thirst, is not dependent upon peripheral changes in the mouth. Contrary to Cannon (1915, 1932), as dry mouth or stomach contractions are not the causes of thirst or hunger, changes in sodium composition of the oral cavity are not the cause of a salt hunger. In other words, while perhaps salivary sodium influences the perception of salt, changes in it are not necessary for the expression of the hunger. Sodium hunger, like other motivated behaviors, is a central motive state (e.g. Stellar, 1954).

PERIPHERAL NERVES AND THE IDENTIFICATION OF SALTINESS

In this section, I call your attention at the outset to two facts: 1) the chorda tympani nerve is especially important in the detection of NaCl; it is an important afferent limb for determining sources of sodium in the world; and 2) there is a specific NaCl–LiCl sensitive channel within the chorda tympani nerve. First, consider the anatomy of the peripheral gustatory nerves.

The seventh, ninth, and tenth cranial nerves convey gustatory information from taste receptors in the oral cavity to the brainstem (e.g. Pfaffmann, 1967,

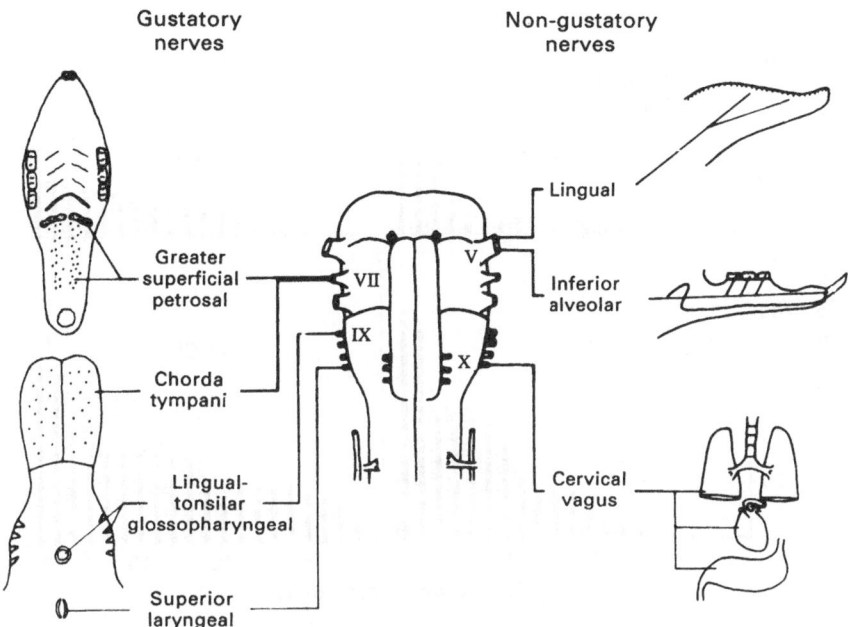

Fig. 3.1. Peripheral nerve description. Diagram of gustatory nerve innervation of the solitary nucleus from the oral cavity (Hamilton & Norgren, unpublished observations).

Fig. 3.1). These cranial nerves have long been known to be involved in ingestive behavior (Herrick, 1905, 1948).

It is the seventh nerve that contains the majority of sodium receptor sites. Perhaps, how salty something is perceived to be is to some extent determined by the activation of this nerve. That is, the salty taste is known to activate the chorda tympani nerve, or seventh cranial nerve (Pfaffmann, 1955). While all regions of the tongue are responsive to gustatory stimuli, the anterior region of the oral cavity is most responsive to salty and sweet ingesta; the posterior end to quinine or bitter ingesta (Pfaffmann, 1967). The acceptance of a salty-tasting source arises in part by the activation of the anterior end of the oral cavity (seventh or chorda tympani nerve), and possibly the inhibition of the posterior region (ninth or glossopharyngeal nerve). Rejection is the converse (Nowlis, 1977; see also Travers *et al.*, 1987). In fact, electrical stimulation of the anterior tongue in decerebrated cats elicits acceptance responses, while stimulation of the posterior tongue elicits rejection responses (Nowlis, 1977).

While the sodium that bathes the oral cavity affects the chorda tympani nerve, or seventh cranial nerve (e.g. Smith *et al.*, 1978), is the chorda tympani nerve sensitivity changed when rats are sodium hungry? When responses from

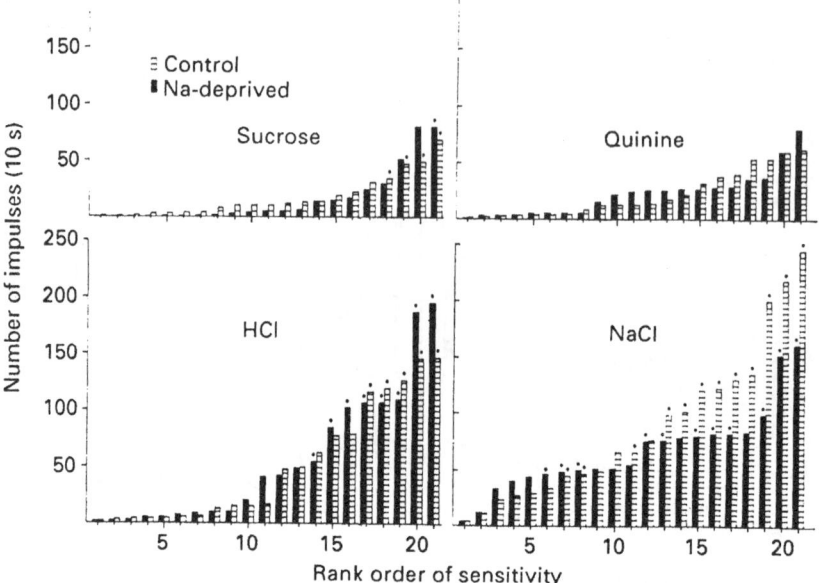

Fig. 3.2. The distribution of responses of 21 fibers in sodium-deprived rats and of 21 fibers in controls to 0.5 M sucrose, 0.02 M quinine hydrochloride, 0.01 M HCl, and 0.1 M NaCl rank ordered (from low to high) four different times according to the size of the responses to each stimulus. Responses size refers to the number of impulses elicited in 10 s of stimulations. A dot above a bar indicates whether a particular fiber is a sucrose-best fiber (*top left*), a quinine-best fiber (*top right*), an HCl-best fiber (*bottom left*), or an NaCl-best fiber (*bottom right*). The chief difference between fibers from sodium-deprived animals and controls is in responses to NaCl (*bottom right*), where fibers which are more sensitive to salt (75 or more impulses elicited in 10 s of stimulation) were smaller in deprived animals' nerves (from Contreras, 1977).

whole nerve chorda tympani were first recorded, no change was found following adrenalectomy (Pfaffmann & Bare, 1950; Nachman & Pfaffmann, 1963). Richter had hypothesized (1956) that sodium hunger resulted in a change in the perception of salt: that its intensity changed. He thought the salt was more intense when it was needed (caution not to confuse changes in intensity with changes in quality). What was later found was not greater sensitivity but less sensitivity. In other words, the saltiness was diminished in intensity when rats needed sodium.

Specifically, salt taste fibers (those fibers most responsive to sodium and lithium salts) within the chorda tympani nerve decrease their firing rate to intraoral infusions of NaCl following sodium deprivation or adrenalectomy

(Contreras, 1977; Contreras, Kosten & Frank, 1984; Fig. 3.2). This occurs at the single fiber level (Contreras, 1977) or whole nerve level (Contreras *et al.*, 1984). Moreover, only salt taste fibers, not fibers responding best to other taste qualities, showed this magnitude of decreased response. There was also a reduced response to LiCl and a smaller effect on KCl (Contreras *et al.*, 1984). This suggests that the reduction takes place in those stimuli that taste salty, such as NaCl, LiCl and, to a much lesser extent, KCl. However, the salt taste ganglion cells are largely responsive to NaCl and LiCl (Boudreau *et al.*, 1983). But, within the chorda tympani nerve, there appear to be fibers that are salt specialists that respond selectively to sodium or lithium salts, and some that are generalists that respond to a variety of salts, including potassium salts (Frank, 1985). None the less, what all these gustatory stimuli have in common is that they taste salty.

There is further evidence that the chorda tympani branch of the seventh cranial nerve, appears to mediate specific salty taste perception. Rats demonstrate an increase in their detection threshold for NaCl following transection of the chorda tympani branch of the seventh cranial nerve (Spector *et al.*, 1990). This fact was revealed using a specially designed 'gustometer'. Water-deprived rats were trained to drink small volumes of water and to suppress the licking response when a solute is present to avoid shock. Detection thresholds were determined by lowering the concentrations of the taste stimuli. Chorda tympani transection increased the NaCl threshold by over two orders of magnitude (Spector *et al.*, 1990); there was no effect on the threshold for sucrose. Moreover, sectioning a branch of the ninth nerve had no effects on NaCl detection thresholds. Chorda tympani section also eliminates NaCl/KCl discrimination in sodium replete rats (Spector & Grill, unpublished observations). These results provide further evidence that anterior gustatory field is important for normal sensitivity to NaCl.

There is, perhaps, further evidence for the role of the seventh nerve in salt taste perception. Chorda tympani-transected rats do not show a shift in oral motor patterns to NaCl when it is infused into the oral cavity over a one-minute period during a taste reactivity test (Schwartz, unpublished observations). They also, do not ingest the sodium to the same degree as intact rats when they are allowed access to it over a 10-min period (Nitabach *et al.*, 1988), and chorda tympani-sectioned sodium depleted rats are less selective than intact rats in this response to sodium and non-sodium salts (Breslin *et al.*, unpublished observations; Fig. 3.3). But, over longer time periods (24 hours), chorda tympani sectioned sodium hungry rats do ingest the salt normally (Pfaffmann, 1952; for review, see Richter, 1956).

The salt taste fibers appear to be served by a sodium-specific transduction mechanism. Note that one can block sodium ions from crossing the dorsal lingual epithelium using topical treatment of amiloride (Heck *et al.*, 1984; Hill

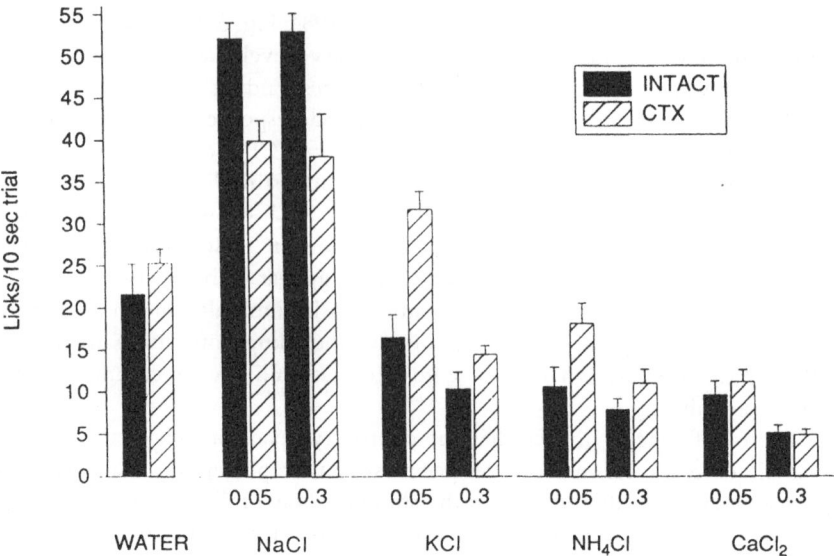

Fig. 3.3. Licking patterns for water and salts in sodium deplete chorda tympani sectioned (hatched bars) and intact rats (Breslin, Spector & Grill, unpublished observations).

& Bour, 1985). Amiloride is a specific blocker for sodium and lithium transport and not for potassium (though in frogs it suppresses both NaCl and KCl, see Yoshi *et al.*, 1986). This blockage of sodium transport eliminates the sodium-specific response of the chorda tympani nerve. It is also known to block LiCl taste responses; this is not surprising given the great physicochemical similarities between Na and Li ions. 'Acid-best' fibers are not affected by the amiloride treatment (Frank, Contreras & Hettinger, 1988).

Amiloride application to the oral cavity in humans also decreases salty taste perception; it also increases sodium transport which results in enhanced salt taste perception (Schiffman *et al.*, 1986). Amiloride is also known to reduce salt intake induced by sodium depletion in rats (Bernstein & Hennessy, 1987), but does not entirely eliminate sodium salt taste perception (Formaker & Hill, 1988). There is a residual chorda tympani response to sodium salts following amiloride pretreatment, but this is probably an effect of the anion.

Thus, taste transduction at the level of the peripheral gustatory nerves plays a functional role in the identification of sodium or lithium ions. Recall that, of all the nonsodium salts, LiCl is the most salty, is actually preferred to a number of sodium salts initially when rats are sodium hungry (Nachman, 1963*b*; Schulkin, 1982; Fig. 3.4) and for which they will run down a runway (Schulkin *et al.*, 1985a). The sodium-hungry animal only stops ingesting the LiCl, presumably, after it begins to become sick from its ingestion (Chapter 1).

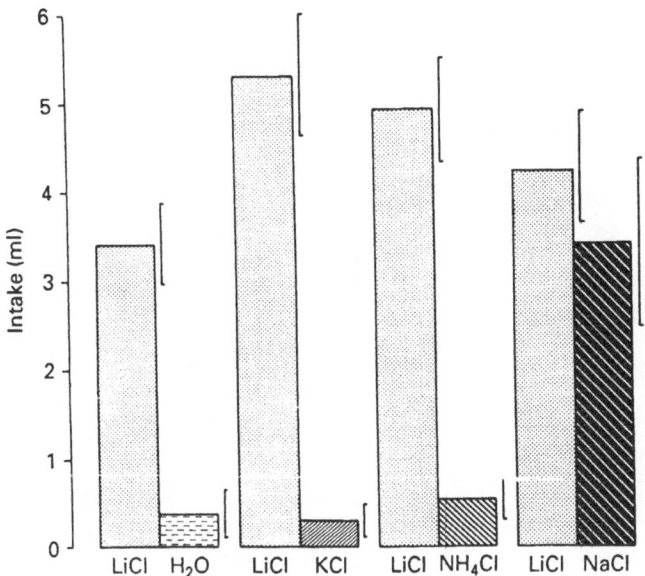

Fig. 3.4. Mean intake of solutions in 15 minutes by adrenalectomized rats in two-bottle preference tests. The salts are all 0.12M solutions. SE of each mean is marked off by vertical lines (from Nachman, 1963*b*).

INTERLUDE: BASIC SALT TASTE CATEGORY

One view of salt taste perception is consistent with the theory that there may be fundamental taste categories (e.g. Bartoshuk, 1980): basic qualities out of which more complex taste sensations are constructed. Salt is one; the others are sweet, bitter and sour.

But, the issue of whether there are such singular taste qualities remains unresolved (Erickson & Covey, 1980; Scott & Chang, 1984). The query of whether some qualities are inherently singular and the basis for synthesizing other taste qualities is rather old and a difficult one to answer. My own Kantian view is that there are, indeed, basic analytic gustatory categories that provide the basis for our gustatory sense of the world. Indeed, this is a suggestion that is beginning to be put to experimental test.

CENTRAL GUSTATORY SITES

Salt appetite is paradigmatic of a taste-guided behavior. It is the central nervous system that ultimately decides whether or not to ingest the salt when sodium hungry. Different levels of the gustatory neural axis participate in the decision. Consider some of the central sites that play a role in salt appetite and probably other taste-guided behaviors.

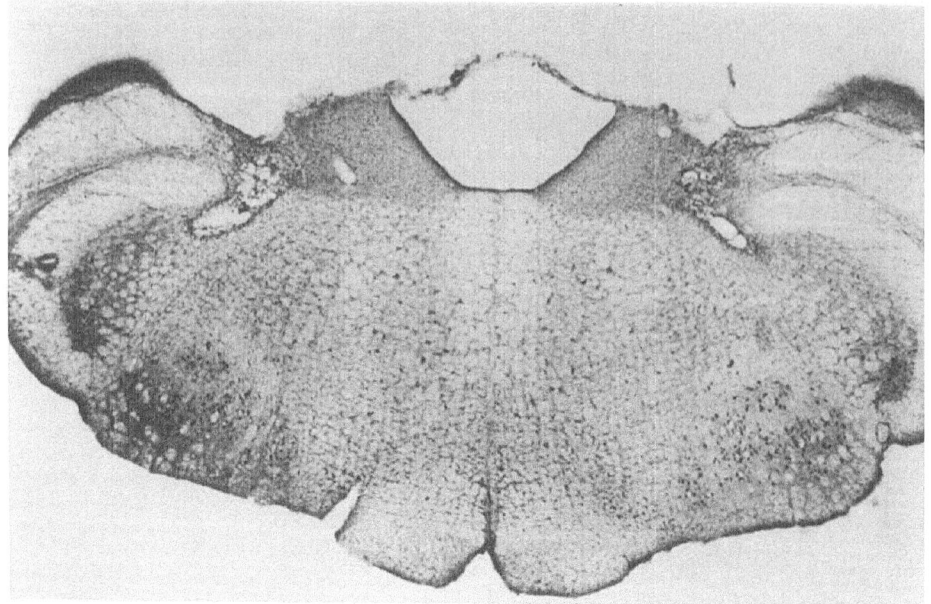

Fig. 3.5. Representative electrophysiologically guided NTS lesion (from Schulkin *et al.*, 1985*b*).

The seventh, ninth and tenth cranial nerves transmit gustatory information to the anterior region of the solitary nucleus (NTS) (Norgren, 1984). From the NTS there is a projection to the medial parabrachial region (PBN). This represents a second gustatory region in the brainstem (Norgren, 1984). Electrophysiologically guided lesions of either the gustatory portions of the NTS or PBN (Fig. 3.5, 3.6) abolish or reduce salt intake following sodium depletion (Fig. 3.7). They also abolish any change in the oral facial profile (described in Chapter 1) following intraoral infusions of NaCl (Schulkin *et al.*, 1985*b*, see Flynn *et al.*, 1992 for more details). These two brainstem gustatory sites, therefore, are essential for the sodium hunger that results from body sodium depletion, which, in the case of the rat, is mediated by the actions of angiotensin and ALDO – the hormones of sodium conservation (Chapters 2, 4 and 5). Both regions may also be involved in other taste-guided behaviors (e.g. taste aversion learning; Flynn *et al.*, 1992).

As will be described in more detail in Chapter 5, decerebrate rats do not ingest sodium when rendered sodium hungry by body sodium depletion (Grill *et al.*, 1986), nor do they show normal taste reactivity to intraoral infusions of NaCl (Flynn & Grill, 1988). Their ingestive response to different concentrations of NaCl is compromised by the decerebration. Note that decerebration

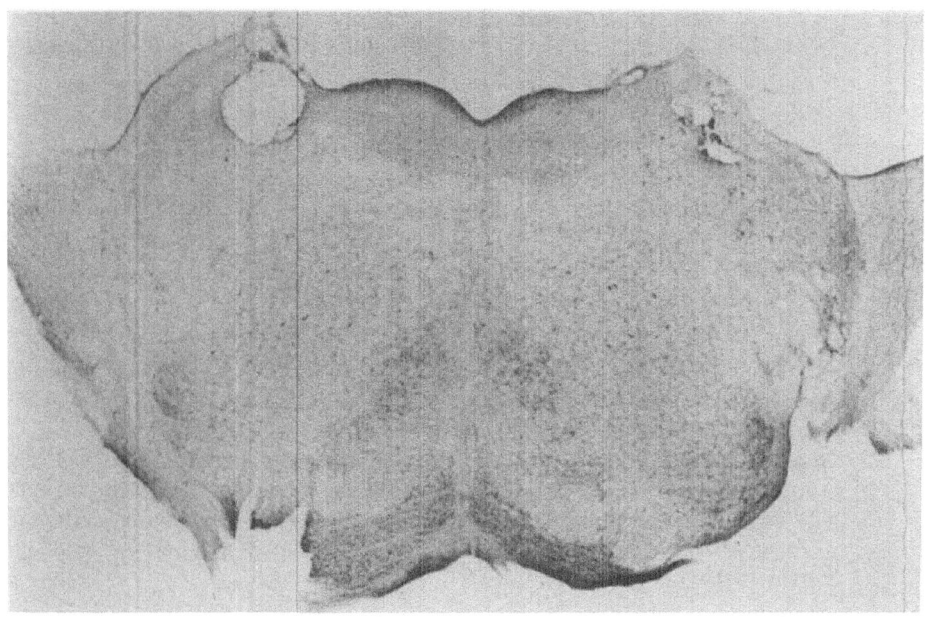

Fig. 3.6. Representative electrophysiologically guided PBN lesion (from Schulkin *et al.*, 1985*b*).

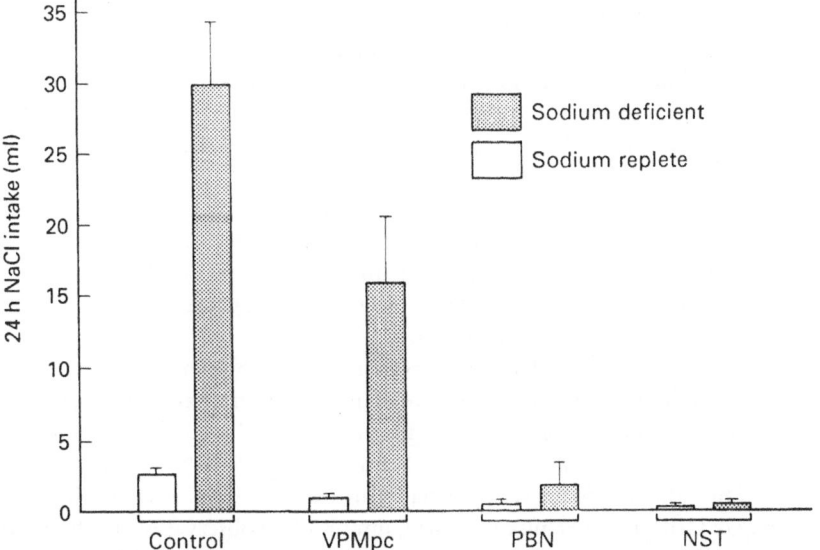

Fig. 3.7. Salt ingestion of lesion and control groups when sodium replete and sodium hungry (from Schulkin *et al.*, 1985*b*).

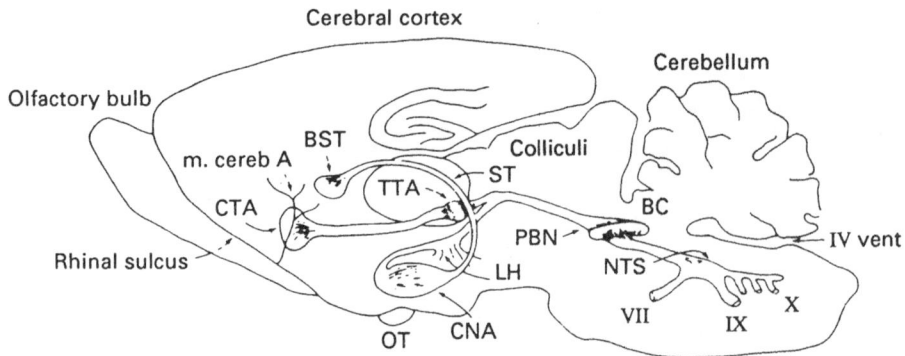

Fig. 3.8. Schematic outline of thalamo-cortical and limbic connections for taste in rat brain. NTS, nucleus of solitary tract; LH, lateral hypothalamus; PBN, parabrachial nucleus; BC, brachium conjunctivum; TTA, thalamic taste area; CTA, cortical taste area; CNA, central nucleus of the amygdala, ST, stria terminalis; BST, bed nucleus of stria terminalis; OT, optic tract; IV vent, fourth ventricle; VII, IX, X, taste afferents of seventh, ninth and tenth cranial nerves; m. cereb, A, middle cerebral artery (from Norgren, 1984).

disconnects forebrain sites from the PBN and NTS which are located in the brainstem. Moreover, it is known that decerebration results in decreased firing patterns of the NTS to intraoral infusions of NaCl (Scott & Mark, 1986). Therefore, this level of the gustatory neural axis appears involved in salt taste perception; but, while the lower brainstem is involved in salt appetite, the forebrain is also essential for the behavior (Chapter 5).

There are two major fiber projections from the PBN in the rat (Norgen 1984; Fig. 3.8; but see Norgren, 1984 for species differences). The first is the classic thalamic–cortical sensory pathway. This is the pathway that may determine the sensory characteristics of a stimulus – whether it is salty or sweet, or sour or bitter (Pfaffmann *et al.*, 1977). This pathway projects to the ventral basal thalamus (VPMpc) which, in turn, projects to the neocortical gustatory area (e.g. Wolf, 1968a, Kosar *et al.*, 1986).

Lesions to either the thalamic taste region itself or the removal of the whole cortex (including the gustatory portion) do not impair the expression of sodium hunger from sodium depletion; nor other regulatory ingestive behaviors, e.g. thirst and hunger (Wolf *et al.*, 1970; Schulkin & Grill, 1980; Schulkin *et al.*, 1985). They also do not impair the latent learning for salt described in Chapter 1 (Wirsig & Grill, 1982; Schulkin & Grill, unpublished observations). Therefore, the dorsal pathway from the PBN to thalamic and neocortical sites is not essential for the expression of salt appetite.

It is the ventral pathway from the PBN to the central amygdala that is important for sodium hunger. This pathway may subserve the rewarding value of the salty, sweet, sour, or bitter tastants (Pfaffmann *et al.*, 1977). It was C. Judson Herrick (1905), the great comparative anatomist, who discovered this ventral projection in the catfish. This projection, which is much denser than the dorsal pathway and which, perhaps, subserves general visceral functions for motivated behavior (Pfaffmann *et al.*, 1977) projects to the lateral hypothalamus, *en route* to the central nucleus of the amygdala and the bed nucleus of the stria terminalis (Norgren, 1984).

There are direct connections from the central nucleus of the amygdala (and extended amygdala – the bed nucleus of the stria terminalis) to the NTS and PBN, and from the NTS and PBN directly to these forebrain sites (e.g. Schwaber *et al.*, 1982). This direct reciprocal connectivity with these brainstem sites may allow for the centrifugal control and organization of acceptance or rejection of ingesta. In fact, as we will see, the central nucleus of the amygdala is known to be involved in sodium ingestive behaviors; damage to this region, as will be discussed in Chapter 5, interferes with the avidity for NaCl in sodium-hungry rats. Thus, this ventral pathway from the PBN and the NTS may, therefore, be important for motivation and reinforcement, suggesting that it may be suited to monitor taste and drive-related signals in the organization of motivated behaviors. In other words, it is the ventral pathway from brainstem gustatory sites that is involved in salt ingestion, in addition to other motivated ingestive behaviors (Stellar & Stellar, 1985). More work needs to be done to understand the role of particular forebrain targets of this gustatory pathway in controlling salt ingestion and other taste-guided motivated behaviors.

GUSTATORY NARROW AND BROAD TUNING

Gustatory sites in the brain are characterized partially in terms of their distributions of narrow and broadly tuned cells (e.g. Scott & Chang, 1984). Gustatory cells become more broadly tuned as one ascends into the forebrain; in other words, peripheral and lower brainstem gustatory sites are more specialized for discriminating the analytic taste qualities mentioned earlier. Therefore, it has been suggested that brainstem sites are more indicative of 'pure' gustatory processing than higher gustatory sites which presumably respond to other stimuli (Scott & Chang, 1984). At the level of the seventh nerve, the NTS, and the PBN, these gustatory cells are, perhaps, more narrowly tuned. At the level of the forebrain, the cells are perhaps more broadly tuned (thalamic, amygdala and gustatory neocortex), responding to a larger variety of gustatory and nongustatory information. The point is that, as one moves up the gustatory neural axis, the taste coding becomes less specialized; the same holds for the perception of something salty (Travers & Smith, 1979).

SALT TASTE DEVELOPMENT

The development of gustatory receptors is essential for the expression of sodium hunger, and other ingestive behaviors that depend upon the gustatory system. Young rats, interestingly, accept NaCl solutions in concentrations rejected by adults (e.g. Bernstein & Courtney, 1987). In addition, lingual application of amiloride has no effect on salt perception in the first days of postnatal development (Hill & Bour, 1985) as it does during adulthood. In the hamster, which shows no preference for salt in adulthood, the responsiveness actually decreases with age (Hill & Mistretta, 1990). The rat's electrophysiological responsiveness in the NTS and the PBN to NaCl infusions is later to develop than that for other gustatory stimuli (Hill *et al.*, 1983; Hill, 1987*a*, *b*). Human infants also initially accept, or are indifferent to, concentrated salt solutions (Desor *et al.*, 1975). Rejection increases with age (Beauchamp *et al.*, 1986). Salt taste magnitude estimates decline with age (Bartoshuk, 1988). But many factors affect salt taste perception in humans including personality (Shepherd & Farleigh, 1986), the amount of salt being consumed (Pangborn & Pecord, 1982), and ethnic backgrounds (Bertino & Chan, 1986). West Indian blacks for example, ingest great amounts of salt; whereas Eskimo Indians ingest little salt (Denton, 1982).

Mothers usually provide their offspring with a surplus of sodium. Under certain conditions, however, the mother may either have an abundance of, or be deprived of, sodium. Exposure to a high sodium diet can increase salt ingestion during adulthood (Contreras & Kosten, 1983). Acutely depleting infant rats of sodium, or depleting the mother of sodium during pregnancy, can increase their sodium intake in adulthood (Nicolaidis *et al.*, 1990). On the other hand, the fact that there is such development in salt taste responsiveness led Hill and his colleagues (and in contradiction to the above results) to the finding that rats would actually ingest less salt in adulthood if deprived of sodium during a critical stage in development (Hill, 1988). This later finding is somewhat analogous to the fact that sensory deprivation of, for example, visual stimuli during critical stages results in decreased visual acuity (Hubel & Weisel, 1979).

HEDONIC PERCEPTION OF SALT

As I stated earlier, P.T. Young emphasized the importance of hedonic factors in food choice in the 1940s and 1950s when narrow behavioristic thinking dominated academic psychology (1949, 1952, 1959). He showed behaviorally how palatability was an important category in the objective study of behavior. And, for the psychophysicist, there has always been an important distinction between the sensory properties of a stimulus and its palatability to the animal (e.g. Bartoshuk, 1979). In Chapter 1, I discussed the fact that the palatability of NaCl and other solutes associated with NaCl is hedonically more pleasing to the sodium-hungry animal (Berridge *et al.*, 1984). This reflects a change in the motivational state of the animal. But, saltiness can be recognized when the animal is either sodium replete or when sodium hungry, so can other taste qualities. The hedonic value attached to the saltiness is what changes when the animal is sodium hungry. It is now valued and pleasing. Evolution has provided a mechanism that ensures that ingesta essential for maintaining homeostasis are hedonically rewarding (Cabanac, 1971, 1979; or Stellar, 1980).

This psychological event is reflected by changes in the central nervous system. Gustatory neurons in the NTS show decreased firing patterns to oral infusions of NaCl following sodium deprivation (Jacobs *et al.*, 1988 – as noted earlier, a similar finding occurs in the chorda tympani nerve, following sodium deprivation, or depletion; though, see Hill, 1988). There is also a change in the firing pattern in the gustatory portion of the NTS of 'sweet-best' or 'reward' neurons. They increase their firing pattern following oral stimulation of sodium when rats are deprived of sodium in their diet over a 10-day period (Fig. 3.9). Bitter and sour-best neurons show little change in firing rate. When sodium is needed, the 'reward neurons' firing patterns are increased almost 1000-fold, and the taste of salt becomes rewarding. The reward neurons in the NTS are, perhaps, importantly involved in the transformation of salt taste perception and its acceptance, or rejection; similar findings in taste aversion learning reveal that sweet-best neurons decrease their firing to specific ingesta following taste aversion learning (Scott & Mark, 1986). In terms of salt, this does not mean that salt tastes sweet to the sodium-hungry animal. Quite the contrary. I believe it still tastes salty. But now it is rewarding to the animal.

Recall that oral infusions of hypertonic NaCl evoke a mixed acceptance–rejection oral–facial profile when sodium replete, while, when sodium hungry, the rats solely exhibit ingestive responses (Chapter 1). The oral–facial ingestive profile looks something like it does when a rewarding stimulus is infused into the oral cavity. The gustatory region of the NTS appears to be part of a circuit that underlies this shift in salt acceptability when rats are sodium hungry. Note, however, that a decerebrate rat, which lacks the latter stages of the ventral taste pathway, does not exhibit a shift in palatability when rendered sodium hungry. (Grill *et al.*, 1986; Chapter 5).

Behaviorally, the approach to, and avoidance of, ingesta is phylogenetically ancient (Schneirla, 1959). The acceptance, and rejection of, ingesta are mediated by two independent systems. Kent Berridge (Berridge & Grill, 1983; Berridge & Schulkin, 1989) and his colleagues have provided the most convincing behavioral evidence for the existence of two independent hedonic evaluation systems. Using the taste reactivity test, they showed that both acceptance and rejection responses often occur at the same time. Figure 3.10 describes a model that accounts for both responses. The hypothesis is that acceptance and rejection responses compete with one another for expression; the palatability judgment is

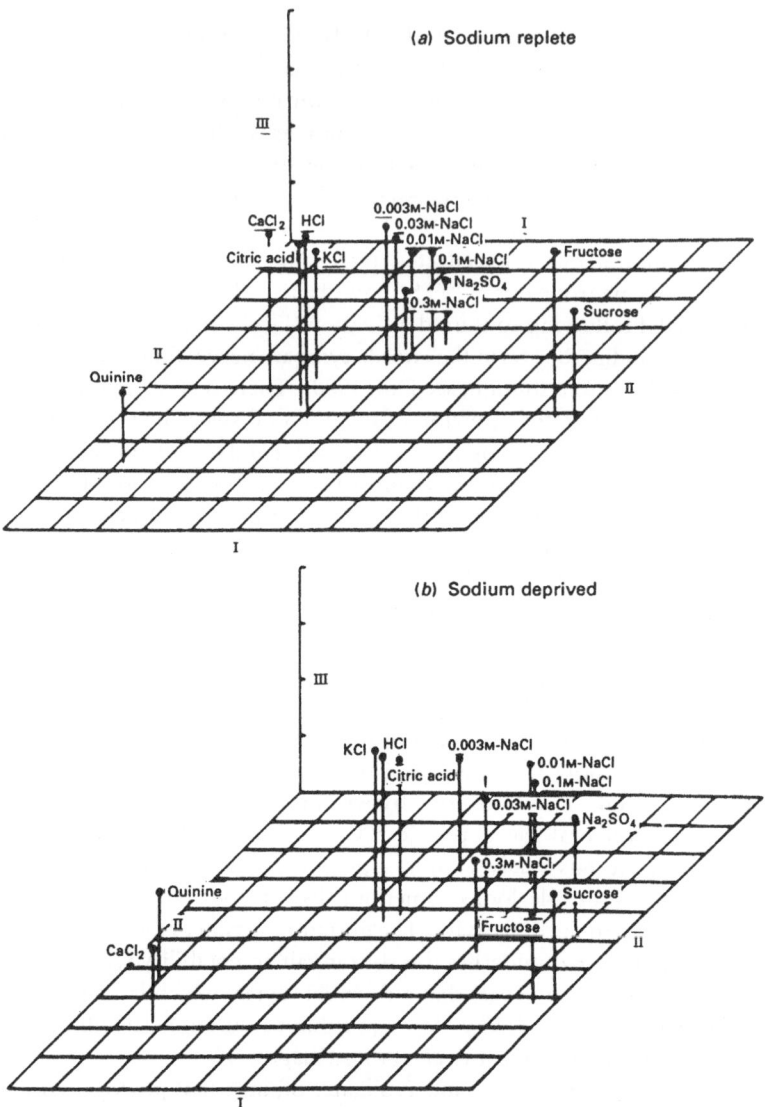

Fig. 3.9. Three-dimensional 'stimulus spaces' derived from inter-stimulus correlation coefficients calculated across four point profiles for each stimulus. Each point was generated by averaging evoked activity in four identifiable neuron types. The space derived from replete rats (*a*) is typical of a normally functioning taste system. A major division is between sweet and nonsweet stimuli. Within nonsweet compounds, distinctions are possible among salts, acids and quinine. The space representing stimulus quality in deprived subjects (*b*) differs in that sodium salts make a closer approach to sugars (from Jacobs, Mark & Scott, 1988).

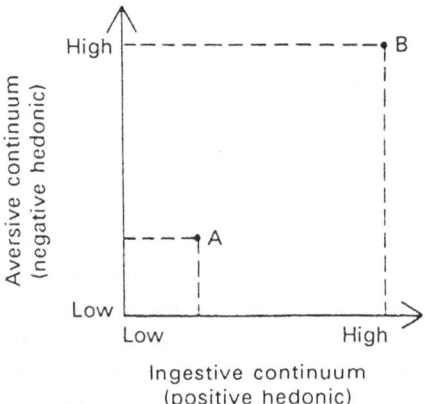

Fig. 3.10. A two-dimensional model of palatability. Point A represents a taste that weakly activates both ingestion and aversion and results in a passive response. Point B represents a taste that strongly activates both ingestion and aversion and results in active alternation of ingestive and aversive consummatory responses. The palatability of other tastes may be represented by points falling anywhere within the plane described by the axes of ingestion and aversion (from Berridge & Grill, 1984).

reflected by the weighted value of the ingestion and rejection response to the intraoral infusions of hypertonic NaCl or other taste stimuli, e.g. citric acid or a sour taste (Berridge & Grill, 1984). Important evidence for the theory of two independent palatability systems, one positive and one negative, is the fact that both expressions can be emitted to varying degrees at the same time: hypertonic NaCl or saltiness, in addition to other qualities, elicits such mixed responses.

By contrast, Young (1959) felt that the expression of both positive and negative responses should not occur. In his view, hedonic experience was on a single continuum and thus positive or negative at any point in time, but not both. But, as I indicated, both positive and negative responses to NaCl infusions can be simultaneously expressed (Berridge *et al.*, 1984). This separation of systems is also expressed anatomically. Recall that there is evidence that the seventh nerve, which innervates the anterior tongue, is involved in acceptance, while the ninth cranial nerve, which innervates the posterior tongue, may be involved in rejection (Nowlis, 1977). At the level of the PBN, infusions of NaCl evoke a mixed inhibitory–excitatory signal in sodium-replete rats (DiLorenzo & Schwartzbaum, 1982), and, at the level of the hypothalamus, stimulation of the lateral region promotes approach to ingesta, and stimulation of the medial region promotes withdrawal from ingesta (Stellar *et al.*, 1979). Rejection and acceptance responses are expressed by the manipulation of separate regions of the amygdala (Fonberg, 1975). Also, at the level of the frontal cortex, there exist separate systems for acceptance and rejection of ingesta; the right hemisphere, known for its involvement in unpleasant events, is responsive to bitter ingesta (quinine); the left is responsive to positive ingesta. This phenomenon is seen in the neonate (Fox & Davidson, 1986; Fox, 1985).

In summary, there is a hedonic shift in the perception of salt that results from sodium hunger; this is reflected in greater ingestive and fewer rejection responses to infusions of hypertonic NaCl. The change in the hedonic value of the NaCl contributes to the motivation to search, and then ingest, the salty substances. The brain regions perhaps involved in the acceptance and rejection of ingesta associated with hedonic events include the seventh and ninth cranial nerves, the NTS or PBN gustatory sites, classical hypothalamic regions, the amygdala and the frontal cortex.

SALT PREFERENCE

There is an important distinction between preference and appetite (Young, 1949). Ingestion and preference are not simply driven by homeostatic needs. Richter almost seems to suggest that they were (see also Cabanac, 1971). Young was quick to point to a difference. Foods or fluids can be preferred, yet not be needed.

The emphasis in this section is on the need-free preference for isotonic saline over water exhibited by several strains of rats. Interestingly, other salts (e.g. KCl) are not preferred over water (e.g. KCl, VanHemel, 1976). There is something special about either the salty taste of dilute NaCl or its postingestive effects. As I said earlier, 'salt'-best fibers in the chorda tympani nerve respond much more to NaCl than to other salts. These fibers could play a role in need-free sodium ingestion.

Eliot Stellar and his colleagues (Weiner & Stellar, 1951; Epstein & Stellar, 1955), for example, showed that water-deprived rats ingest greater amounts of dilute NaCl in 1 h than water or other concentrations of salty water (Fig. 3.11). Sodium-depleted rats also prefer, and ingest, greater amounts of the dilute NaCl than water or other concentrations of NaCl. The authors suggested that the preference for these concentrations of NaCl is not based on learning, since it was manifested within minutes upon their first exposure to the NaCl solutions.

But, while rats prefer NaCl in water, they do not when it is mixed with food (Beauchamp & Bertino, 1985). However, sodium-hungry rats, as noted in Chapter 1, will regulate their salt intake whether in fluid or food. None the less, it has always struck me as interesting that the concentration of NaCl that the rat prefers is the concentration of sodium that bathes the organs of the body, about 0.9%. This holds for both need-induced appetite and need-free preference (e.g. Bare, 1949). Of course, the ideal solution to drink would then be at, or near, this

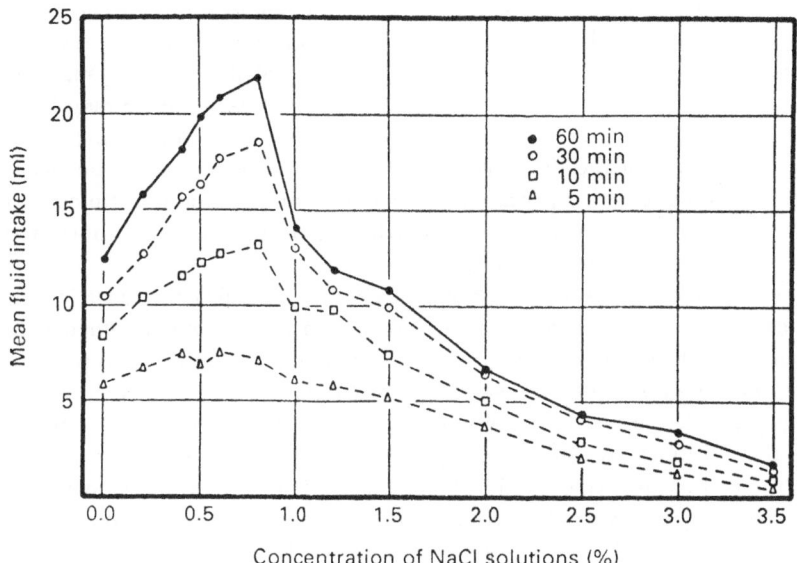

Fig. 3.11. Salt preference-aversion functions based upon fractions of 1-h. Test periods with different concentrations of salt solution. The data are combined from two independent ascending series of salt concentrations (from Weiner & Stellar, 1951).

concentration, the concentration of extracellular fluid. In fact, when offered a choice between water and 0.15 M NaCl, rats begin to ingest the salt more rapidly than if it is hypertonic when they are rendered hypovolemic or sodium hungry (Smith & Stricker, 1969). To repair extracellular fluid deficits, the animal must regulate both water and sodium (Chapter 4); the ingestion of this isotonic concentration makes sense. The fact that Nature made salt preferable at this concentration would fit this view.

But is this NaCl preference, manifested in some species of rats, anything like a sugar preference? I think not. NaCl preference as an hedonic event has been challenged (Deutsch & Jones, 1960). First, it was suggested on the basis of electrophysiological evidence that the ingestion of isotonic or hypotonic saline solutions that are ingested in greater amounts than water is due to a decrease in electrophysiological signals of the dilute water–salt (Deutsch & Jones, 1960). Rats drink more of the dilute NaCl because there is a diminished electrophysiological signal. Second, rats that are thirsty, and which manifest an inverted U-shaped preference–aversion function for saline, do not run for it down an alleyway when not thirsty (Schulkin *et al.*, 1985a, as described in Chapter 1, also cf. Young & Falk, 1956). It is unlike the preference for something sweet. There, the rat will still run down an alleyway to acquire it or bar press in a Skinner Box for it when not hungry (Stellar & Stellar, 1985). Thus, the preference for dilute NaCl will not result in the rat working or running for it when not in some state of alteration of body fluid or sodium balance. The rewarding aspect of the dilute NaCl and the preference for it do not seem strong. This suggests that perhaps need-free sodium ingestion is not mediated by the same brain mechanisms as those which generate the highly motivated behavior of the sodium–hungry rat.

The rewarding aspect of dilute NaCl may not reflect oral factors at all. In fact, one can see the preference–aversion function in some contexts when taste stimulation is held constant by directly manipulating infusions of sodium into the stomach (Mook, 1963; also see Borer, 1968). By contrast, concentrations of dilute NaCl that are orally preferred are not preferred when water is infused into the GI tract (Mook, 1963); therefore, some extraoral receptor site determines this ingestion of the dilute NaCl. This is in contrast with the consummatory behavior of the sodium-hungry rat, which, in its early stages, is mainly taste guided. Also, consider that adrenalectomized rats ingest the sodium, despite the fact that water is infused into the stomach (Mook, 1969). In this context, the sodium hunger results in a palatability change, whether the salt satisfies the need or not. That is, in this case, the innate hunger for salty commodities overrides all other factors and the animal desperately ingests the salt, despite the fact that it is not rewarded for it.

Interestingly, NaCl seems not to taste good or bad when initially infused into the oral cavity. Recall, that the taste of NaCl at hypertonic concentrations is not initially rejected by the rat, when it is infused into the oral cavity. This effect is concentration dependent with the higher concentration eliciting rejection faster than the lower concentrations (Schwartz & Grill, 1984). This decrease in salt intake, perhaps like taste aversion, is probably due to postabsorptive mechanisms. There are two likely physiological candidates: the stomach and the liver. Evidence in rats (Stellar *et al.*, 1954; McCleary, 1953) and in dogs (Chernigovsky, 1962) suggests that intragastric intubation of NaCl decreases NaCl preference. Moderately thirsty rats, for example, intubated with different concentrations of NaCl, ingest greater or lesser amounts of the NaCl depending upon the concentration of NaCl intubated and the concentration of NaCl that is offered to the animal.

There is compelling evidence that the need-free saline preference itself may be due to the activation of hepatic sites. Hepatic portal infusions of isotonic NaCl decreases dilute NaCl intake in thirsty rats (Blake & Lin, 1978). Infusions of water do not have this effect. These effects are abolished following cervical vagotomy. Therefore, the vagus nerve, or tenth cranial nerve, known to transmit visceral information to the brain, also appears to be conveying information about consequences of salt ingestion. Candidate sites in the brain for gustatory and hepatic and/or stomach interactions point to the PBN (Rogers *et al.*, 1979). Infusions into the hepatic portal vein, of NaCl, activate gustatory sites within the PBN (Chapters 4 and 5). We will see later that hepatic receptors, known for their involvement in the regulation of food intake (for review, see Friedman, 1982), also play a role in satiation of need-induced sodium hunger.

Also, consider that central dopamine and the endorphins increase NaCl preference behavior (Gilbert & Cooper, 1987; Kuta *et al.*, 1984), as does damage to different regions of the amygdala (Gentil *et al.*, 1968). Moreover, electrical stimulation of the amygdala either increases or decreases need-free salt ingestion; septal stimulation only decreases salt intake (e.g. Grace, 1968). In addition, lateral hypothalamic stimulation increases salt intake (McKenzie and Denton, 1974). Stimulation of the anterior cingulate cortex decreases salt intake, while stimulation of what the author has called the 'prelimbic area' increases salt intake (Chiaraviglio, 1984).

SPECIES DIFFERENCES AND SALT INGESTION

There are interesting species differences in the ingestion of salt to note for the comparative psychobiologist. Not all mammals respond to salt taste. Cats and dogs do not have units which respond specifically to NaCl and LiCl in the geniculate ganglion of the chorda tympani nerve, while rats, sheep and goats do (Beidler, 1954; Boudreau *et al.*, 1985). Also, it is difficult, though not impossible, to demonstrate sodium hunger in the carnivore (Chapter 1), which expresses little need-free preference for NaCl (Carpenter, 1956).

Interestingly, both pigeons and chickens, and even laughing gulls demonstrate a preference for isotonic saline (Duncan, 1962; Harriman, 1967), and it is known that the pigeon demonstrates a hunger for sodium when given natriorexigenic treatments (Chapters 1 and 2). Maybe, if challenged, the other two birds would as well. Perhaps having a preference for NaCl is a precondition for expressing a hunger for sodium when it is needed. This is an intriguing idea that would be nice if it were true, but it is not. Gerbils, for example, have no preference for NaCl but will ingest it if they need it (Chapter 1). Strains of rats (Fisher) that do not prefer NaCl when mature (Bernstein, 1988) will ingest it when sodium is needed (e.g. Midkiff *et al.*, 1985). In addition, they will express a positive shift of their oral–facial profile to infusions of NaCl (Grill & Bernstein, 1988). On the other hand, those strains of rats (Sprague–Dawley) that have a need-free preference for salty water ingest more salt than they need, when made hungry for it, while those who do not have the need-free preference do not and tend to ingest what they need (Rowland & Fregly, 1988*a*).

Sheep, on the other hand, and other herbivorous mammals that have no elevated need-free ingestion of salty water do not act as salt gluttons when hungry for it; they ingest the salt commensurate with what they need. Interestingly, rabbits express a large *ad libitum* preference for KCl solutions and a small one for NaCl solutions (Contreras *et al.*, 1985; see also Carpenter, 1956). Rabbits also ingest NaCl commensurate with their sodium deficit (Denton, 1982). Mice of different strains respond differently to NaCl solutions (Hoshishima *et al.*, 1962), but they generally have a blunted preference for NaCl (Wolf & Lawrence, 1963), and drink about as much sodium as they need to repair their deficits (cf. Denton *et al.*, 1988; Rowland & Fregly, 1988*c*). Hamsters in which it can be difficult to elicit a salt appetite (Rowland, 1986) show no preference for salt (Carpenter, 1956). This evidence is somewhat supportive for the idea that a preference for NaCl is a precondition for ingesting excessive amounts of it when sodium hungry, but the evidence is clearly equivocal.

GUSTATION AND SALT SATIETY

Physiological psychologists have, for many years, been interested in the role of the mouth and postabsorptive factors in the satiation of hunger and thirst (e.g. see Miller's collected works, 1971*a*, *b*). Thirst or hunger, for example, was more readily satiated when water or food were ingested than when water or nutrients were intragastricly intubated. Gustation has long been thought to be the primary factor in terminating the appetite for salt. Early studies in both sodium-depleted sheep and rats found little or no satiation of salt intake following intragastric intubation (within 10 to 30 minutes following intubation); there was satiation when sodium ingestion was allowed (e.g. Nachman & Valentino, 1966; Bell *et al.*, 1981). But, with longer-term infusions there is a clear reduction of salt intake (Kissileff & Hoeffer, 1975). Moreover, it was also observed that, when there were longer intervals between the time of the intubation and the salt appetite test, there was a clear reduction in the salt intake of sodium–depleted rats (Levy & McCutcheon, 1974; Wolf *et al.*, 1984; Fig. 3.12). In fact, one can see a clear reduction of salt intake within 30 minutes following intragastric intubation of NaCl, but not other solutes or substances (Wolf *et al.*, 1984). Therefore, at this time, the reduction in intake is due to the sodium.

Of course, the oral ingestion of salt compared with the intragastric route is more satiating, as is clear from Fig. 3.12. However, under different conditions and a longer time frame (90 minutes), when one compares the amount of NaCl ingested in rats either allowed to predrink NaCl or infused in equal amounts with the NaCl in the hepatic portal circulation, the intakes are comparable (Tordoff *et al.*, 1987; see Chapters 4 and 5). Additionally, sodium–depleted sheep, will bar press for sodium for an intravenous infusion, and they can also be sated when sodium is infused into the cerebral ventricles (as discussed in Chapters 4 and 5). Therefore, the oral ingestion can be bypassed and sodium

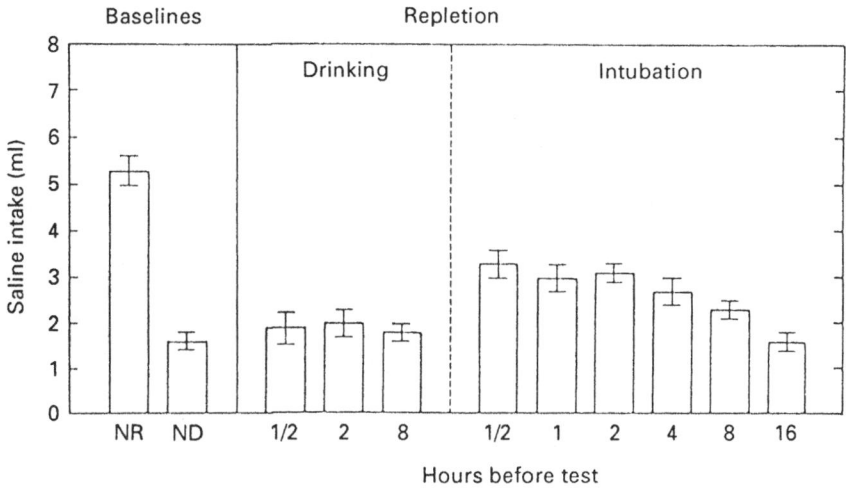

Fig. 3.12. Salt intake of sodium–depleted rats at different times following ingestion or intragastric intubation of sodium. NR, sodium replete; ND, sodium deplete (from Wolf, Schulkin & Simpson, 1984).

hunger can still be reduced. But, when minimal amounts of sodium are absorbed, following intragastric cannulation (Tordoff *et al.*, 1987; Denton, 1982) both sodium–depleted sheep and rats ingest large amounts of sodium (Fig. 3.13). That is, oral factors alone are insufficent to terminate the hunger for sodium at normal levels of intake. A similar phenomenon holds for water-deprived sheep or rats (Denton, 1982; Blass & Hall, 1976), or hungry rats drinking sweet solutions (e.g. Sclafani & Nissenbaum, 1985).

In each of the above cases of salt reduction in the absence of oral stimulation, gustatory sites may still be activated. That is, the intragastric, hepatic, intravenous and even the infusion of sodium into the ventricles may activate central gustatory sites and thereby decrease salt intake. As noted above, we know that hepatic portal infusions of sodium activate central gustatory sites (Hermann *et al.*, 1983). Also, denervation of the liver directly affects salt taste perception (Deems & Friedman, 1988*a*, *b*). Intravenous infusions can activate hypothalamic sites which are also activated by oral infusions (Nicolaidis, 1969). Therefore, central gustatory sites are informed of the sodium absorption which contributes to the decrease in the salt ingestion of the sodium-depleted animal.

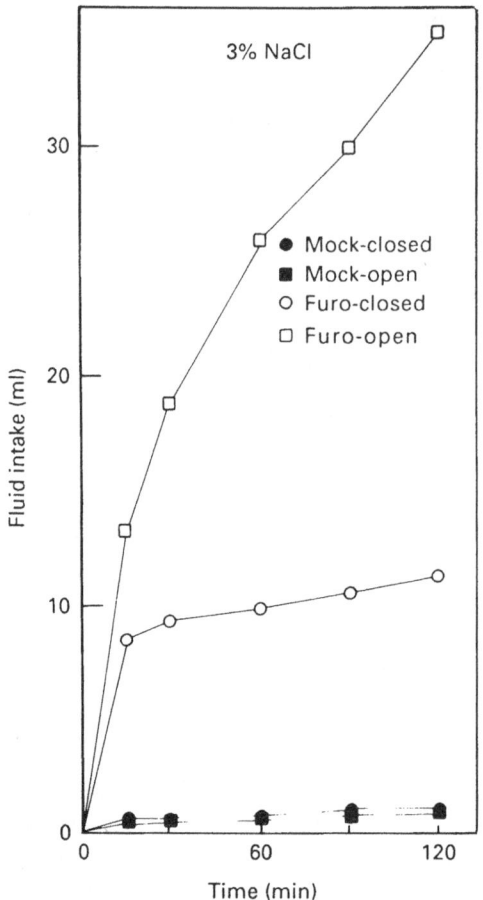

Fig. 3.13. Intake of salt solution when sodium hungry or sodium replete with gastric cannulas open or closed (from Tordoff, Schulkin & Friedman, 1987).

CEPHALIC INFLUENCES

Cephalic influences play an important role in the digestion of food; insulin secretion to food intake is perhaps a good example (Powley, 1977). In addition, oral stimulation in body fluid-deprived dogs is known to reduce ADH secretion within minutes (Thrasher *et al.*, 1982), and is known to affect the electrophysiological responses in a variety of ventral forebrain sites involved in the regulation of body fluid homeostasis (Nicolaidis & Jeulin, 1986). These are the same hypothalamic sites where gustatory afferents terminate (Norgren, 1970),

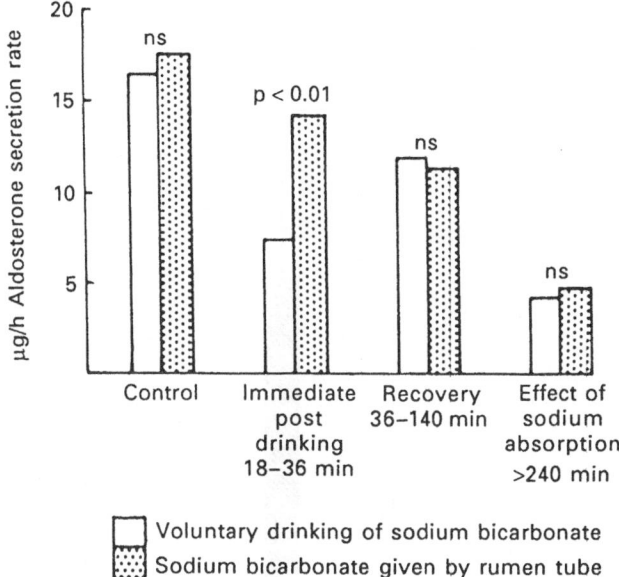

Fig. 3.14. Comparison of the effect on mean aldosterone secretion rate of correction of sodium deficiency by voluntary drinking of sodium bicarbonate solution with the effect of administering approximately the same amount of sodium bicarbonate into the rumen by tube. The aldosterone secretion rate at 18–36 min following voluntary satiation of salt appetite was significantly less (P < 0.01) than with rumen tube (from Denton, 1982).

and which respond to systemic infusions of hypertonic NaCl or oral stimulation of the NaCl (Nicolaidis & Jeulin, 1986).

Cephalic-driven regulatory changes also occur after tasting salt, for example, when sodium-deficient, aldosterone secretion is reduced within 20 min after the ingestion of sodium (Denton, 1982, Fig. 3.14). When sodium is delivered by gavage, aldosterone levels take much longer to go down.

SEARCHING FOR SOMETHING SALTY
Salty commodities are noticeable, and animals search for something salty when they need sodium (Schulkin, 1982). While the evidence points to a specific appetite for sodium, sodium-deficient rats will ingest KCl (e.g. Falk, 1965a; Jalowiec et al., 1966), and LiCl (e.g. Nachman, 1963b; Schulkin, 1982). These compounds share a taste salty to humans (Bartoshuk, 1980; Schulkin, 1982). In fact, sodium-hungry rats tend to prefer the saltiest tasting salts (Schulkin, 1982; Fig. 3.15). Saccharine or nonnutritive sweets (like nonsodium salts) are perhaps

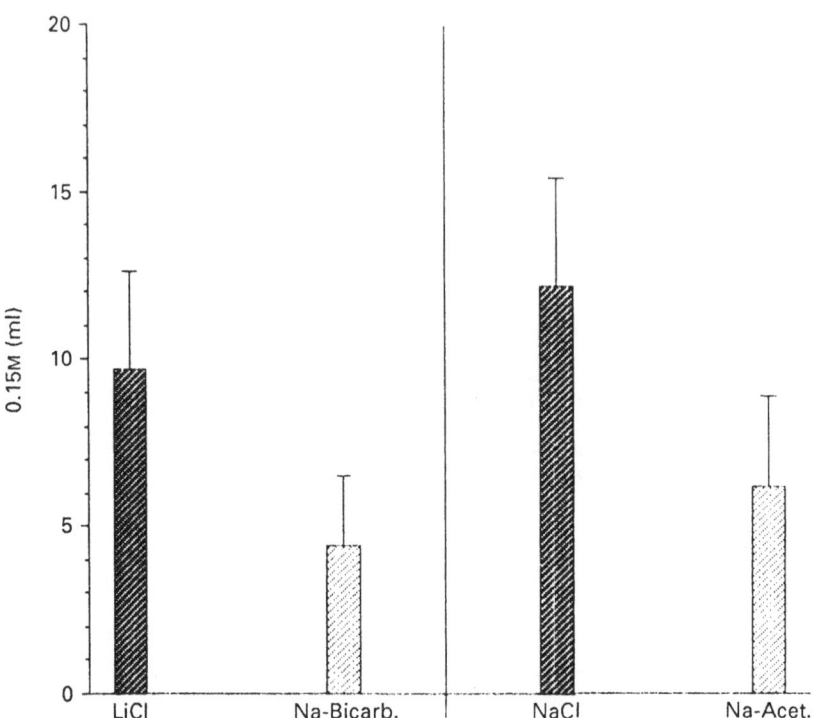

Fig. 3.15. The 2-h ingestion of salt solutions (LiCl vs Na-bicarbonate) or (NaCl vs Na-acetate) over a 2-h period. Both the NaCl or the LiCl when judged by human subjects as the saltier tasting salts of the pair (Schulkin, unpublished observations, 1982).

also ingested in hungry rats because they signal sources of nourishment (LeMagnen, 1985).

It should be noted that lithium chloride and NaCl taste very much the same to people, though they can be discriminated (e.g. Schulkin, 1982). The taste generalization of NaCl to LiCl in rats is evidence that the two ingesta taste similar (Nachman, 1963a), and, in rats, the ingestion of LiCl (because of the gastrointestinal sickness that results) decreases NaCl ingestion to a much greater extent than other solutes. But, unlike other taste aversions, when the rat is sodium hungry, it will overcome the aversion and ingest the salt (e.g. Stricker & Wilson, 1970; Trent & Kalat, 1977). This occurs regardless of earlier preferences and aversions that can influence later salt ingestion, when sodium-deficient rats go for sodium (cf. Harriman, 1955; Cullen, 1969; Grimsley, 1970).

Interestingly, other mineral deficiencies such as potassium (Adam & Dawborn, 1972; Fig. 3.16), or calcium (Tordoff *et al.*, 1990, Fig. 3.17) result in

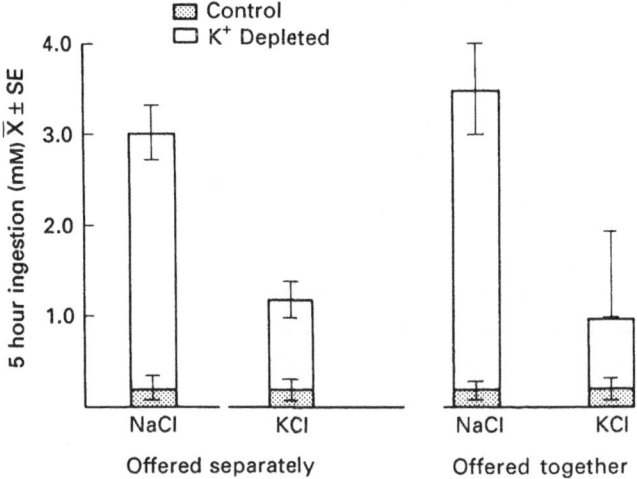

Fig. 3.16. The 5-h ingestion (means and SEM) of 0.67M NaCl and 0.4 M KCl when offered separately or together in potassium–depleted and control rats (from Adam & Dawborn, 1972).

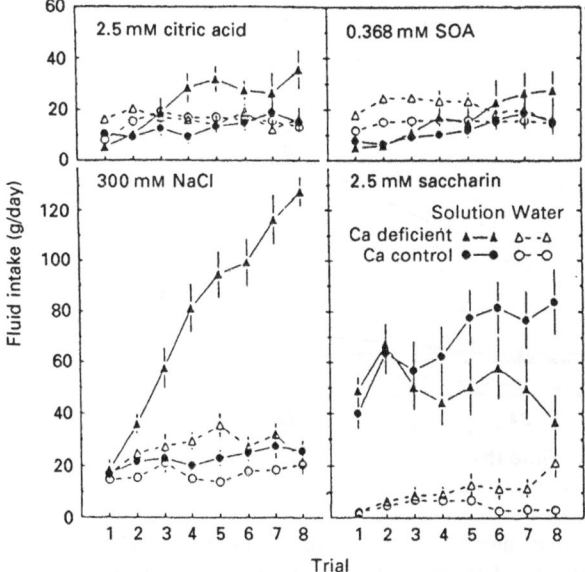

Fig. 3.17. Intake of water and tastants by rats fed calcium deficient or control diets over a 32 day period. COA, sycrose octaacetate (from Tordoff, Ulrich & Schulkin, 1990).

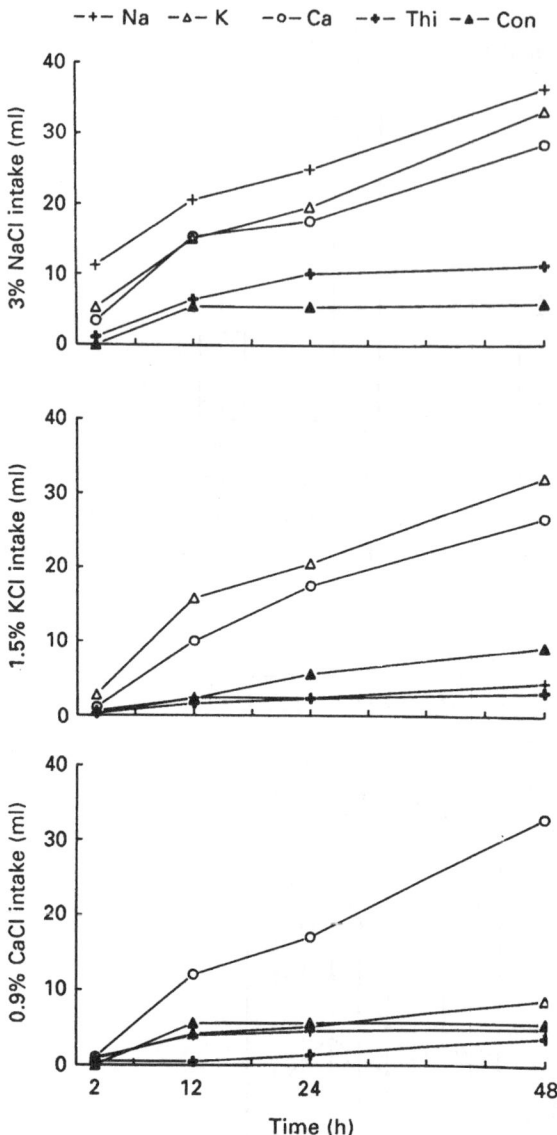

Fig. 3.18. Ingestion of various minerals in young rats (about 45 days of age) who were placed on either a sodium-deficient, potassium-deficient, calcium-deficient, thiamine-deficient or control diet. Rats were generally on the diet for 2 weeks, and then given access to the salt solutions. Intakes were monitored at 2, 12, 24 and 48 hours (from Schulkin, 1986).

Fig. 3.19. Elephants at a salt mine in South Africa (from Ian Redmond).

NaCl ingestion. Rats eventually ingest the mineral that is needed, but continue to ingest the NaCl (at least in one study as can be seen from Fig. 3.18). Vitamin-deficient rats do not behave this way (Schulkin, 1986). Calcium-deprived rats will even bar press for sodium in addition to calcium salts (Lewis, 1968). In humans, iron deficiency can result in 'salt pica' (Shapiro & Linas, 1985).

Recall from Chapter 1 that a variety of mammals and birds are attracted to the salt licks which are rich in sodium in addition to other minerals (e.g. Herbert & Cowan, 1971). Figure 3.19 depicts a herd of elephants in a salt mine. The male elephants are generally pushed out of the herd when they reach maturity. It is therefore the female, perhaps, that orients the herd to the salt lick. This search behavior may not necessarily be related to low sodium in the diet. But, at the salt licks, their mineral requirements are met. This is particularly true during reproduction when the female has more need for the minerals. Sodium can act as a signal for the presence of other needed minerals. This then allows the innate sodium recognition system to be used in the search for these other minerals.

It is interesting that NaCl is the prototypical salty taste (Bartoshuk, 1980; Schiffman, 1980). All other sodium or nonsodium salts either taste less salty or have additional taste properties, e.g. chalky and bitter. The fact that, in some instances, sodium-hungry rats will actually ingest nonsodium salts over sodium salts, or choose to ingest more salty-tasting sodium salts over less salty tasting sodium salts (e.g. Falk, 1965a), suggests that it is the salty taste they are searching to ingest when they are sodium hungry. There is some evidence that humans will grind up potassium and ingest it when sodium deprived (Denton, 1982); possibly it is for the salty taste, not for the potassium. The foods and fluids we ingest, of course, have more than one taste. In fact, many salts have more than one taste. As shown earlier, NaCl is the pure salty taste. The ingestion of this salt which mostly activates the salt taste fibers, activates other taste fibers as well, just not as much (Frank *et al.*, 1983).

The specialist–generalist gustatory fiber distinction parallels one that Paul Rozin made for ingestive strategies (1976). Some animals are exclusively generalists. The omnivorous rat or person eats lots of things. The specialist *par excellence* is the herbivore, which eats only specific kinds of leaves. The specific hunger for salty commodites is an example of a specialized response in the restoration of an imbalance. Vitamins, and other mineral deficits, require general learning about either averting the *malaise* of the deficiency, or to a lesser extent, feeling the well-being of restoration of the needed vitamin or mineral before specific preferences develop (Rozin, 1976). While rats may not have an innate specific appetite for zinc, calcium, potassium or magnesium (Rozin & Schulkin, 1990), the specialist sodium system is used to solve other mineral requirements; that is, the salience of the specialized salt taste plays a role in the regulation in several of other mineral deficiencies, e.g. calcium or potassium, and perhaps even zinc (Jakinovich & Osborn, 1981). A possibility is that one strategy which evolution has selected for, in the behavioral regulation of minerals, is the ingestion of salty tasting commodities. The salty taste in mineral licks is a tag for minerals; thus, the animal is able to restore what is needed, and excrete what is not, when it licks. The innate hunger for sodium orients animals towards the lick by the salience of the salty taste and the desire and pleasure of its ingestion.

CONCLUSIONS

In this chapter, various conceptual and empirical observations have been presented regarding the role played by the gustatory system in the detection and ingestion of salt. At each step, one could see how the behavior, physiology and anatomy are interwoven. In the next chapter, we continue looking at the integration of behavior, physiology and anatomy, and we add the molecular and the pathological domains.

4   Physiological factors in the control of salt intake

INTRODUCTORY

This chapter is in the tradition of 'whole body physiology'. This tradition, like that of physiological psychology (e.g. Stellar et al., 1954) emphasizes behavioral control of regulatory events. In this chapter, the reader will get a glimpse of whole body physiology at work.

A number of organs are involved in the regulation of body sodium and extracellular fluid balance, ranging from the heart, adrenal gland, kidney, and gastrointestinal organs, to the pituitary gland and brain. Moreover, sites of angiotensin or ALDO action, in addition to sodium and volume receptors, suggest common points of interaction in the genesis of a sodium hunger. That is, specific mechanisms exist to regulate body fluids: for intracellular fluids, there are osmotic controls; for extracellular fluids, endocrine mechanisms are essential in the conservation of sodium. These same endocrine mechanisms are also important for the behavioral responses. The mechanisms of sodium hunger are intimately tied to those involved in body fluid balance and cardiovascular regulation. The basic research is therefore of medical importance.

KIDNEY AND BRAIN: THE RENIN–ANGIOTENSIN SYSTEM

Of our body weight, 60% is water, 30% of which is extracellular and 70% intracellular (Ramsay & Ganong, 1977). Hypovolemia, or loss of extracellular fluid volume, results in the ingestion of both water and sodium (Stricker & Wolf, 1969) to help restore extracellular fluid volume.

Sodium homeostatic mechanisms are activated by deviations in body sodium composition. The kidney reduces its excretion of sodium when the body is low in it (Denton, 1982) and fecal concentrations also decrease (Grace et al, 1979). Body sodium and extracellular fluid depletion result in the release of renin from the kidney (e.g. Blair-West et al., 1967; Stricker et al., 1979; Fig. 4.1). Renin acts upon circulating angiotensinogen (which is decreased or increased depending upon the sodium status of the animal, e.g. Iwao et al., 1988) that is released from the liver to form angiotensin I. Enzymes that have their highest concentration in

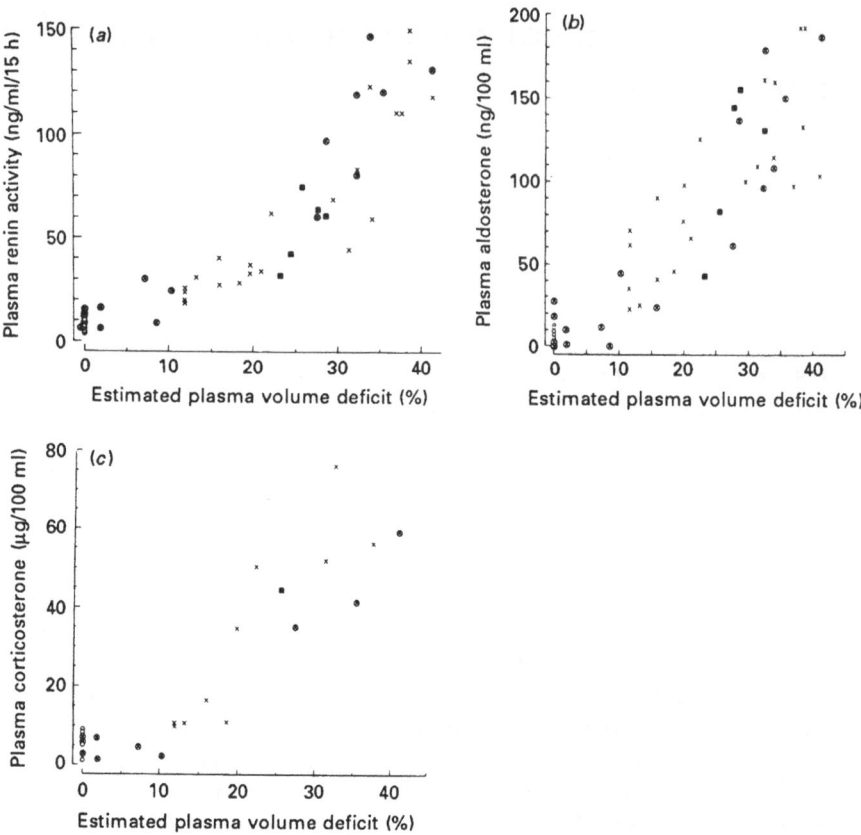

Fig. 4.1. Plasma renin activity (*a*), aldosterone concentration (*b*), and corticosterone concentration (*c*), as a function of estimated plasma volume deficit (from Stricker *et al.*, 1979, and for more details).

the lung convert angiotensin I to angiotensin II, which is the active form of the hormone (Fig. 4.2). Angiotensin increases ALDO release into the blood system (e.g. Aguilera & Cait, 1983), and evokes pressor and cardiovascular responses (e.g. Mimran *et al.*, 1974).

While angiotensin is a primary hormone both behaviorally and physiologically in the regulation of extracellular regulation (Fitzsimons, 1979), nephrectomy does not abolish drinking in response to extracellular fluid depletion (Fitzsimons, 1961; Stricker, 1966). It does abolish the salt intake in response to hypovolemia or extracellular fluid deprivation (Fitzsimons & Stricker, 1971; Chiaraviglio and Taleisnik, 1969). While the peripheral release of renin results in increased drinking behavior, it does not increase salt ingestion (see

Fitzsimons, 1979). Moreover, the following further evidence suggests that it is angiotensin of central, but not peripheral, origin that is important for the appetite in the rat. 1) Adrenalectomy-induced salt appetite is dependent upon angiotensin, because, if its actions are blocked in the brain, then the appetite is abolished (Sakai *et al.*, 1989). 2) And, while nephrectomy abolishes sodium hunger to sodium depletion, intracerebrally delivered angiotensin restores the appetite (Chiaraviglio, 1976). 3) Peripheral blockade of the renin–angiotensin system does not affect salt appetite, but central blockage, of course, does (Sakai & Epstein, 1990a). 4) Angiotensin receptors in the brain are elevated following natriorexigenic treatments (Chapter 5). 5) The synergistic effects of angiotensin and ALDO occur when angiotensin is infused into the brain directly and not when it is given systemically (Sakai & Epstein, 1990a). Taken together, the above results suggest that it is central angiotensin that is the primary factor in the elicitation of salt appetite.

As discussed in Chapter 2, while earlier investigations suggested that elevated systemic angiotensin does not result in salt drinking (e.g. Fitzsimons & Stricker, 1971), later investigations found that salt intake was elevated by systemic infusions of angiotensin (Findlay & Epstein, 1980). But, this appetite in rats was secondary to the sodium loss that resulted (Findlay & Epstein, 1980). By contrast, intracerebral infusions of angiotensin, however, cause a modest salt appetite (Chapter 2), the onset of which actually occurs prior to the sodium loss that the hormone provokes over time (Fluharty & Manaker, 1983). It, therefore, suggests that the salt appetite can be aroused independent of the sodium loss. The same holds for the intracerebral injections of renin. For example, the salt appetite that results from such renin injections appears within 1 h, while the rats are in positive sodium balance (DeLuca Jr *et al.*, unpublished observations).

The case is different for the herbivorous sheep. In this context, peripheral angiotensin is importantly involved in the onset of the appetite; centrally delivered angiotensin elicits a salt appetite which is largely due to renal sodium loss (Weisinger *et al.*, 1987a, b). Thus, while the rat drinks more sodium with systemically delivered angiotensin because of increased sodium loss, the reverse is true of the sheep. That is, it is renally derived angiotensin that seems to be prominent and not centrally generated angiotensin in the elicitation of sodium intake in the sheep; its effects are primarily related to the actions of angiotensin independent of its effects on sodium loss. Moreover, blockade of central angiotensin in sodium-depleted sheep does not affect salt appetite, while peripheral blockade does (Weisinger *et al.*, 1987a, b). The reverse is true for the rat.

There are independent angiotensin systems: a peripheral system that activates ALDO release and is involved in the physiological regulation of fluid volume control, and a central angiotensin system that is importantly involved in the

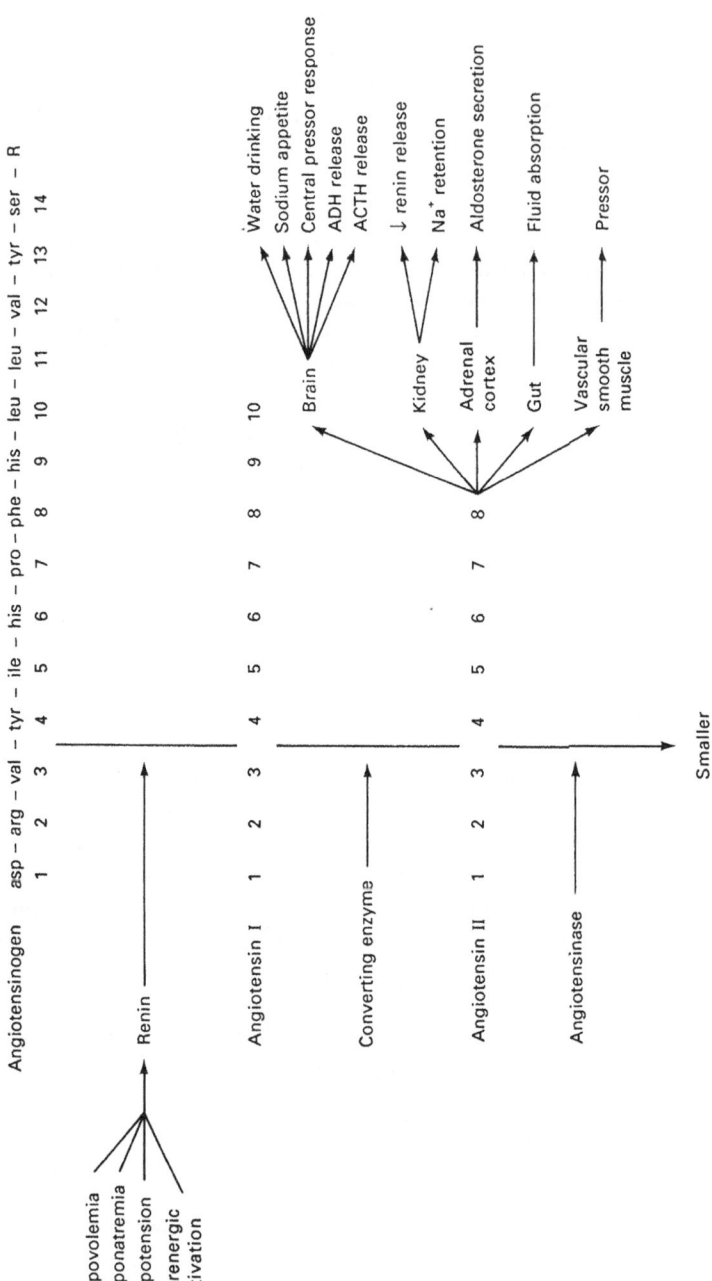

Fig. 4.2. Steps in the conversion of renin to angiotensin and the various physiological and behavioral events (from Epstein, unpublished observations).

elicitation of sodium hunger at least in the rat. But, also note that there are direct projections from the kidney to the solitary nucleus of the brainstem (Wyss & Donovan, 1984), and regions of the solitary nucleus are known to be involved in body sodium homeostasis (Chapter 5). In addition, angiotensin receptors are found in this brainstem region, and in regions known to be involved in cardiovascular and body fluid regulation (e.g. parabrachial nucleus, see Chapter 5). In Chapter 5, evidence will be presented that ventral forebrain sites which synthesize angiotensin, and which are involved in cardiovascular and body fluid regulation, in and around the third ventricle, are involved in the arousal of a sodium hunger.

Despite the species differences, angiotensin for both the rat and the sheep may serve as a brain neurotransmitter in the organization of salt-drinking behavior (Denton, 1982; Chapter 5). In other words, angiotensin plays a dual role in the maintenance of extracellular and sodium balance by acting to promote physiological changes so as to maintain sodium balance and to promote the behavior of salt ingestion.

ADRENALS – MINERALOCORTICOIDS

The mineralocorticoids are the other primary hormones besides angiotensin involved in the maintenance of extracellular fluid volume and sodium balance and, as mentioned earlier, they generate the behavior of salt ingestion. Figure 4.3 depicts the roles of angiotensin and ALDO in the behavioral regulation of intravascular or extracellular fluid volume (Fisher & Buggy, 1975).

The action of ALDO during sodium deprivation and extracellular fluid loss is to redistribute sodium in the body, and to conserve it from loss from the kidney, as well as to absorb it from the parotid gland and the gut (Blair-West *et al.*, 1963; Stricker *et al.*, 1979). A number of factors that are normally associated with sodium deprivation increase ALDO secretion (for review, see Denton, 1982), including ACTH, renin–angiotensin, and high levels of potassium. Physiologically, ALDO is known to affect vasomotor activity as well as to cause increased fluid exchange and pressor effects. While the half-life of ALDO in plasma is about 30 min their sodium-retaining effects appear by 1 h in rats adrenalectomized 24 h earlier (Horisberger & Diezi, 1983). The mineralocorticoid hormones can also increase sodium levels in target tissue (Green *et al.*, 1948), and, in fact, enlarge the tissue (e.g. heart and kidney, Green *et al.*, 1952). Note that serum levels of ALDO are generally higher in females than in males (Yao & Epstein, unpublished observations), and could help explain the enhanced avidity for salt in females (Chapter 2) and their lower excretion of sodium in response to natriuretic treatments (Wolf, 1982).

Increased perfusion of ALDO in the kidney results in what has been called 'mineralocorticoid-escape'. That is, chronic mineralocorticoid treatment results in diuresis and natriuresis that usually occurs after several days. Animals

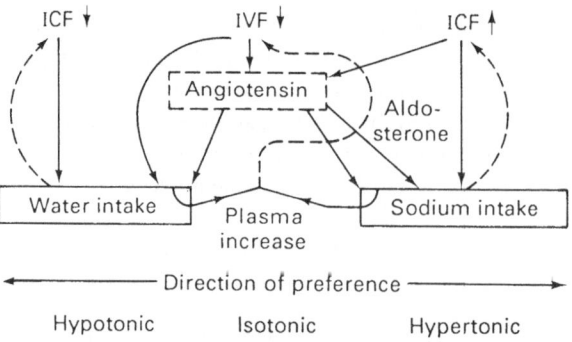

Fig. 4.3. Model of the central control of thirst indicating angiotensin effect on both water and sodium intake. ICF = intracellular fluid decrease (and/or Na+ decrease) in monitor neurons; IVF = intravascular fluid volume (and/or pressure decrease; solid line and arrow = facilitatory effect; dashed line and open arrow = inhibitory (negative feedback) effect (from Fisher & Buggy, 1975).

start to excrete sodium instead of retaining it, perhaps because of its known pressure diuresis (Green *et al.*, 1952), and the release of atrial natriuretic factor (ANF) (Kelly & Nelson, 1987) to help maintain proper body fluid state.

Mineralocorticoid receptors are found in many regions of the body. Within the gastrointestinal tract, these regions include the liver, stomach, duodenum, jejunum, ileum and colon; and the hormone also binds to the lung, aorta, spleen, heart and bone, in addition to the pituitary, kidney and the salivary, parotid and mammary glands (for review, see Funder, 1986). In the brain, the hormone binds to neurons in addition to glial cells (Beaumont *et al.*, 1987).

The binding for mineralocorticoids in renal, as well as brain, nuclei has been reported to go up some sevenfold in rats placed on a sodium-deficient diet (DeNicola *et al.*, 1982) Adrenalectomy is known to increase ALDO receptor sites in the kidney which are then reduced with pharmacological infusions of the hormone (Claire *et al.*, 1981). This induction of ALDO receptors by adrenalectomy is also seen in the brain and colon; potassium consumption reverses this effect (DeNicola *et al.*, 1982).

The principal classes of adrenal steroid hormones, mineralocorticoid, and glucocorticoid hormones such as corticosterone (CORT), compete for many of the same receptor sites in the brain as indicated in Chapter 2 (Funder, 1986). It has been suggested that there are two types of CORT receptors, one of which is responsive to ALDO as well as to CORT (type 1) receptors (DeKloet *et al.*, 1986).

But, there is preferential binding of ALDO over CORT that is clear at the level of the kidney and at many other target sites, in addition to cell culture preparations (Lan *et al.*, 1981). One can see this preferential binding for ALDO in the pituitary gland (Krozowski & Funder, 1981), and in brain tissue (discussed in Chapter 5, McEwen *et al.*, 1986). In the brain, for example, hippocampal binding is more selective for CORT, while that of the preoptic region is more selective for ALDO (Chapter 5: Yongue & Roy, 1987; Coirini *et al.*, 1988). But, note that both types of receptor sites have a high affinity for CORT (McEwen, 1989).

Interestingly, CORT increases ALDO binding (Brinton & McEwen, 1988; Coirini *et al.*, 1988; Chapter 5). When adrenal and glucocorticoid hormones are combined, one sees elevated salt intake (Chapter 2). Recall that the glucocorticoid hormones by themselves are not natriorexigenic in the rat (Wolf, 1965), but in the rabbit they are (Chapter 2).

What the reader should realize is that glucocorticoid hormones circulate in larger amounts than mineralocorticoid hormones (McEwen, 1989). There are also many more glucocorticoid receptor sites in the brain than mineralocorticoid receptor sites (McEwen, 1989). It is also the case that, during sodium depletion, both mineralocorticoid and glucocorticoid levels (in addition to renin–angiotensin) are elevated (Stricker *et al.*, 1979; as in Fig. 4.1).

Glucocorticoids are known for their contribution to energy homeostasis, but are also important in the search for salt (Coirini *et al.*, 1988; Chapter 2) by increasing mineralocorticoid binding in brain regions known to be involved in the hunger for sodium (preoptic–hypothalamic region). This may occur because CORT-preferring sites become ALDO-preferring. The enzyme 11 β-hydroxy steroid dehydrogenase makes this possible because it excludes CORT from ALDO receptor sites (Funder, 1986). The enzyme, in effect, protects the receptor site from CORT (Edwards *et al.*, 1988) which should serve to increase the appetite as the result of the mineralocorticoid hormones.

Steroid hormones, including ALDO, bind to cytoplasmic proteins, and then promote DNA-dependent RNA synthesis of proteins and enzymes involved in sodium transport (Crabbe, 1961; Edelman *et al.*, 1963). Protein synthesis inhibition blocks these effects on sodium transport, which require the activity of ATP (Edelman, 1978). Moreover, these effects are known to occur through amiloride-sensitive sodium transport channels (Will *et al.*, 1980, 1985). Mineralocorticoids are known to affect second messenger systems in the brain (Harrelson & McEwen, 1987), and are also known to affect brain intracellular sodium (Woodbury & Koch, 1957). Figure 4.4 depicts the general model of steroid action on cells (McEwen, 1976). The induction of these processes is known to take hours, which could explain the long latency of sodium hunger emerging when the mineralocorticoids are at nadir levels. Similarly, it takes hours for estrogen to induce lordosis behavior (McEwen & Pfaff, 1985).

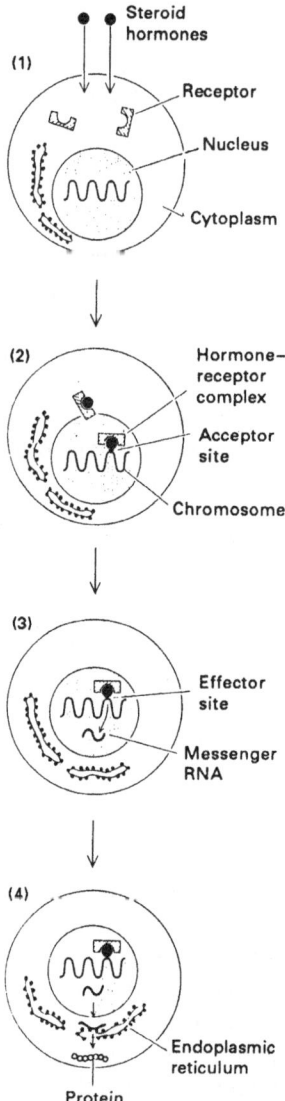

Fig. 4.4. Cellular response to steroid hormones is depicted schematically. Although steroids are presumably able to enter all cells, only certain target cells contain specific receptor proteins that bind the hormone (1). Receptor–hormone complexes are able to pass through the membrane that surrounds the cell nucleus and attach themselves to acceptor sites of the chromosome containing the genes (DNA) of the cell (2). The receptor triggers the synthesis of messenger RNA corresponding to the gene at an effector site (3), which may or may not be different from the acceptor site. Messenger RNA leaves nucleus and migrates to endoplasmic reticulum, where it presides over the synthesis of protein encoded by gene (4) (from McEwen, 1976).

CIRCADIAN ORGANIZATION OF THE RENIN–ANGIOTENSIN–ALDOSTERONE SYSTEM

In Chapter 1, I described how the circadian biological clock is used to anticipate the availability of salt. Sodium excretion is, to some extent, also under circadian control (Mann *et al.*, 1976). The hormones of sodium homeostasis are also under circadian control. There are, for example, higher levels of angiotensin and ALDO in the rat during the dark hours than during light (Hilfehaus, 1976). This is, of course, when the rat would be ingesting salt, in addition to drinking water and eating.

Mineralocorticoid receptors in the brain are also under circadian control. The lowest levels, in this case, are apparent at the onset of the dark phase of the cycle (Reul *et al.*, 1987; Chao *et al.*, 1989). Levels of mRNA are not changed (Chao *et al.*, 1989). This lack of change at the level of the mRNA suggests that the mechanisms responsible for the available receptor change are at a posttranscriptional level.

PITUITARY

The pituitary–adrenal axis is involved in the elicitation of sodium hunger. ACTH, as shown in Chapter 2, elicits a sodium hunger in several species. Also, the activation of the pituitary–adrenal axis is perhaps related to the sodium hunger that emerges during duress, fight, and flight. ACTH promotes the synthesis and release of mineralocorticoid and the glucocorticoid hormones.

But, an intact pituitary gland is not essential for sodium hunger. Hypophysectomized (HYPOX) rats and sheep express a salt appetite in response to body sodium depletion (see Denton, 1982), although their response is compromised (Jalowiec *et al.*, 1970; Schulkin *et al.*, 1989*b*; Fig. 4.5).

Physiologically, HYPOX rats have been reported to have lower levels of angiotensin and ALDO than normally circulate after acute body fluids changes (Balment *et al.*, 1986). None the less, levels of both hormones are sufficiently increased to retain sodium. And, basal levels of the two hormones are normal, while urinary excretion of sodium induced by them is decreased following sodium deprivation (Fregly & Rowland, 1989). HYPOX rats also have decreased natriuretic responses, because of oxytocin's and vasopressin's role in normal sodium excretion (Balment *et al.*, 1986).

Angiotensin receptors in both brain and the periphery (Douglas *et al.*, 1978; Castren & Saavedra, 1989) and mRNA for angiontensinogen (Schulkin *et al.*, 1989*b*; Fig. 4.6) are decreased in HYPOX rats. While the HYPOX rat demonstrates normal mineralocorticoid-induced salt appetite, its salt appetite in response to angiotensin-induced salt intake is compromised (Schulkin *et al.*, 1989*b*). Thus, the decreased ingestion following body sodium depletion may result from decreased functional activity of the renin–angiotensin system in the

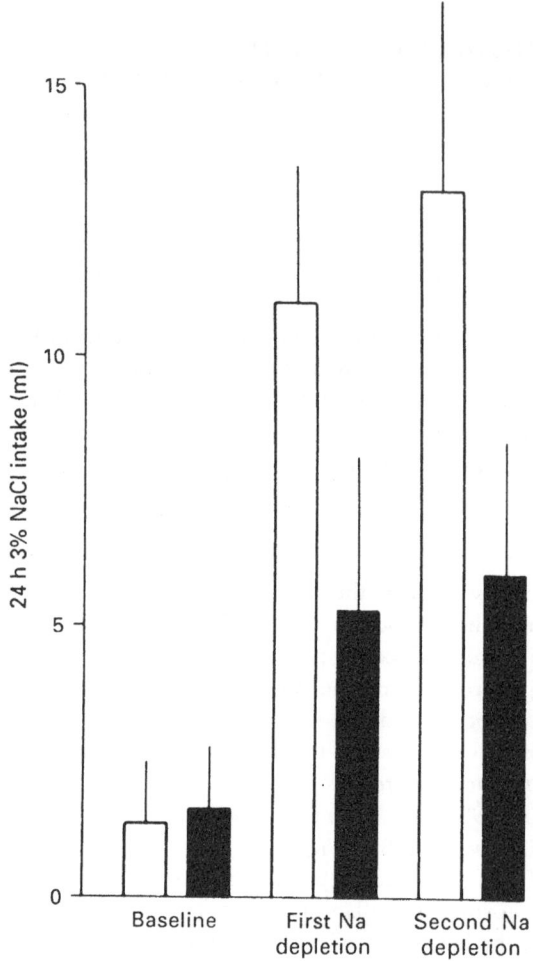

Fig. 4.5. Effect of sodium depletion on the salt intake of HYPOX (shaded) and control rats (from Schulkin, Angulo, Sakai & McEwen, 1989).

HYPOX rat. In addition, the salt appetite that is potentiated by several days of dietary sodium deficiency followed by hypovolemia is decreased in HYPOX rats (Stricker, 1983). This presumably occurs because of the decrease in circulating ALDO which prepares the brain to ingest the salt perhaps by upregulating angiotensin sites. Thus, while the pituitary gland plays an important role in the behavioral expression of body sodium regulation, its importance is tied to the angiotensinergic signal since the HYPOX rat is unresponsive to angiotensin but is responsive to ALDO.

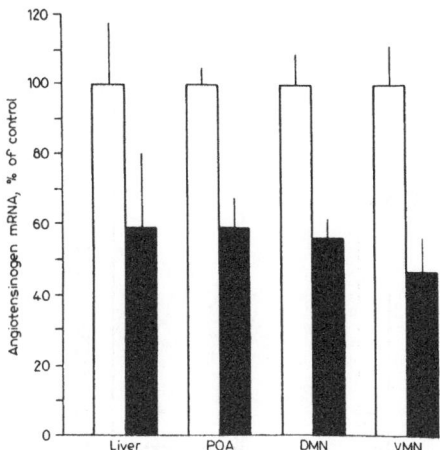

Fig. 4.6. Relative decreases in angiotensinogen mRNA in three brain regions: the preoptic area (POA), the dorsomedial nucleus (DMN) and the ventromedial nucleus (VMN), and in the liver of HYPOX rats (shaded) and controls (from Schulkin, Angulo, Sakai & McEwen, 1989).

HEART: VOLUME RECEPTORS AND ATRIAL NATRIURETIC PEPTIDE

The heart is an endocrine organ (Cantin & Genest, 1986). It releases a peptide hormone that regulates body fluid balance. It is well known that the heart, the organ of cardiovascular control, is essential for body fluid regulation, and is specifically involved in the control of extracellular fluid regulation. It contains pressure or volume receptors that monitor the amount of blood that perfuses it. But the endocrine cells of the heart also release ANF that regulates the volume that passes through the heart and the body (e.g. Genest, 1987). When there is too much volume perfusing the heart, this hormone that is synthesized in the atrium is released to promote sodium and extracellular fluid secretion.

Atrial natriuretic factor is triggered by a number of factors, such as extracellular volume or sodium loading (Genest, 1987). Rapid infusions of sodium through the atrium of the heart promotes ANF synthesis and its subsequent actions on the kidney (Verburg *et al.*, 1986) where its effects are to promote sodium loss. Atrial damage abolishes this response. Infusions of ANF in the systemic circulation or in the cerebral ventricles (Israel & Barbella, 1986; Fitts *et al.*, 1985*a*), are known to promote water and sodium loss (Fitts *et al.*, 1985*a*), and modulate pressor and blood responses (Sills *et al.*, 1985). In addition, eating a salty meal can promote ANF release (Verburg *et al.*, 1986). Atrial appendectomy reduces these effects (Veress & Sonnenberg, 1984). Importantly,

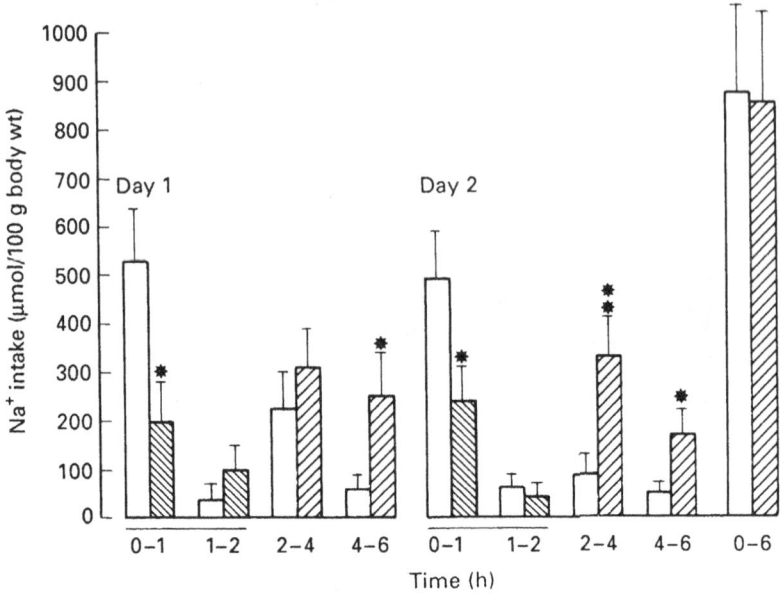

Fig. 4.7. Sodium intake (2.7% saline) of rats that have extracellular fluid deficits produced by peritoneal dialysis with hyperoncotic colloid. Right atrial balloons were inflated in experimental group (hatched bars, $n = 11$) for first 2 h (control group, open bars, $n = 13$). Horizontal line indicates period of balloon inflation. Total sodium intake (0–6 h) for day 1 is shown on extreme right (from Toth, Stelfox & Kaufman, 1987).

ANF decreases the steroidogenesis of ALDO, or angiotensin secretion (Atarashi *et al.*, 1984; Vari *et al.*, 1986), and reduces angiotensin- or ACTH-induced ALDO secretion (Mulrow *et al.*, 1987), and hemorrhage-induced ADH secretion (Samson, 1985). At the molecular level of analysis, ANF also influences second messenger systems by increasing cyclic GMP content, possibly resulting from the stimulation of guanylate cyclase. This occurs in both peripheral and central sites (Israel *et al.*, 1988).

Behaviorally, expansion of plasma volume in sheep, however, does not decrease their sodium hunger to sodium deficiency (see Denton, 1982). In addition, an increase in extracellular volume is not associated with the onset or termination of salt appetite in either rats or sheep (e.g. Denton, 1982; Stricker & Wolf, 1966). None the less, the atrial volume receptors perhaps play some modulatory role in the regulation of water and salt intake. For example, crushing the left atrial appendage abolishes hypovolemic-induced water drinking in sheep, without affecting dehydration-induced thirst (Zimmerman *et al.*, 1981). Decreasing venous return to the heart elevates both water intake in the rat (Fitzsimons, 1979) and water and salt intake in the dog (e.g. Fitzsimons & Moore-Gillon, 1980). An inflatable balloon implanted in the right atrium, at the juncture of the right superior vena cava, reduces hypovolemic-induced thirst, while not affecting dehydration-induced thirst (Kaufman, 1984). When blood pressure is controlled for, ANF decreases the water intake of hypovolemic rats (Kaufman & Monkton, 1988). In addition, distension of the right atrium causes the release of ANF, which appears to be locally mediated (Kaufman, 1984; Meikle & Kaufman, 1988).

And, expanding the balloon is also known to decrease the salt appetite of hypovolemic and DOCA-treated rats (Toth *et al.*, 1987; Fig. 4.7). However, peripheral as opposed to central ANF does not reduce the salt appetite to body sodium depletion (Fitts *et al.*, 1985*a*; Tarjan *et al.*, 1988, Chapters 2 and 5).

In summary, perhaps it is only central ANF that inhibits the appetite. But, it is the expansion of atrial volume receptors that decreases salt intake, and it is the peripheral release of ANF that has important physiological control of body sodium. Perhaps the activation of the volume receptors and the release of ANF act via the vagus nerve on the brainstem to inhibit salt intake, somewhat analogous to gut peptides acting to inhibit food intake via the vagus nerve (e.g. Gibbs *et al.*, 1973). Likely candidates include the area postrema and the solitary nucleus (see Chapter 5).

LIVER

The liver is involved in energy homeostasis and food intake (Friedman, 1982). Hepatic portal infusions of nutrients, for example, decrease food intake (Tordoff & Friedman, 1986). The liver is also important for sodium homeostasis and salt satiety (Chapter 3). It also produces angiotensinogen on which renin acts to produce angiotensin I, and then metabolizes ALDO. Both sodium and volume receptors in the liver are activated by changes in hepatic circulation which also affect cardiovascular functions (Lautt, 1983). Rats that are depleted of sodium have elevated angiotensin receptors in the liver (Sernia *et al.*, 1985). It also appears that both mineralocorticoid and glucocorticoid hormones bind to the liver (Ben-Ari & Garrison, 1988; Duval & Funder, 1974), and that both regulate angiotensinogen mRNA in the liver (Ben-Ari & Garrison, 1988).

Behaviorally, hepatic portal infusions of sodium decrease saline preference in thirsty rats (Chapter 3; Lin & Blake, 1971) and decrease the salt intake of sodium-depleted rats (Tordoff *et al.*, 1986; Fig. 4.8). Other solutes that are infused into the hepatic portal vein, or volume alone, tend not to decrease the salt intake. Infusions of sodium into the vena cava also do not decrease salt intake of sodium-depleted rats (Tordoff *et al.*, 1986); therefore, the decrease in salt intake may be due to activation of hepatic sodium sensors activated by the levels of sodium that circulate in the hepatic portal vein, and not just in the systemic circulation.

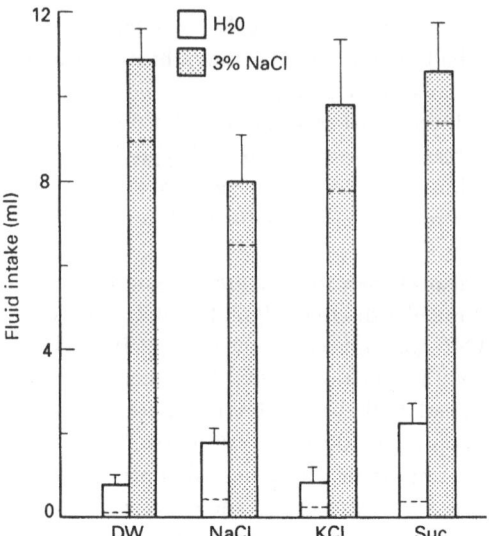

Fig. 4.8. Influence of hepatic–portal infusion of 1 ml/30 min distilled water (DW), or 1.2 osm NaCl, KCl, or sucrose (Suc) on consumption of water and 3% NaCl by sodium-deficient rats. Horizontal striped lines separate intake during last 15 min of infusion from intake during following 15 min (from Tordoff, Schulkin & Friedman, 1986).

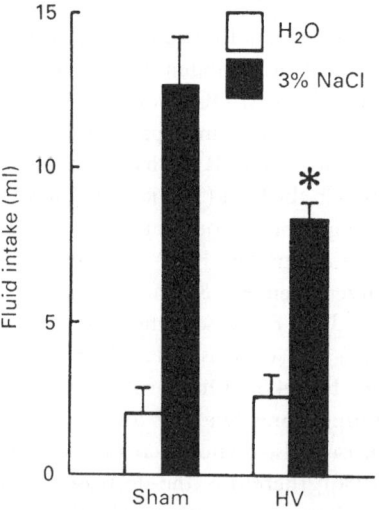

Fig. 4.9. Consumption of water and 3% NaCl by sodium-deficient rats with sham or hepatic vagotomy (HV) (from Tordoff, Schulkin & Friedman, 1986).

Vagotomized rats show diminished dipsogenic responses to extracellular depletions, or to angiotensin (Jerome & Smith, 1982; Martin & Novin, 1981). However, hepatic vagotomy does not decrease ALDO or angiotensin levels to sodium depletion (Tordoff, Fluharty & Schulkin, unpublished observations). But hepatic vagotomy decreases salt intake to body sodium deprivation (Contreras & Kosten, 1981) and from body sodium depletion (Tordoff *et al.*, 1986; Fig. 4.9). Sodium excretion tended to be about the same in the two groups (Tordoff *et al.*, 1986), though hepatic denervation can affect urinary excretion (Levy & Wexler, 1987).

Recall from Chapter 3 that hepatic portal infusions of NaCl activate central gustatory sites at various levels of the neural axis, and activate the hypothalamic–neurohypophyseal system. Neurons in the supraoptic nucleus, for example, are activated and ADH is released from local neurons following hepatic portal infusions (Baertschi & Vallet, 1981). The mechanism of this action is cholinergic and calcium dependent (Stoppini & Baertschi, 1984). In Chapter 5, some evidence will be discussed that cholinergic action in the brain decreases salt intake in sodium-depleted rats.

SODIUM RESERVOIR
Calcium in the bone is a paradigmatic example of a body reservoir. Hormonal signals (Vitamin D and parathyroid) act on calcium bone reservoirs to maintain extracellular calcium balance. Perhaps, a similar mechanism is at work for sodium. Consider the background.

When rats are sodium depleted by a variety of techniques and maintained on Purina chow (which is rich in sodium), they usually demonstrate the appetite for salt some 8 to 24 h later (Wolf & Stricker, 1967; Stricker & Wolf, 1969). Why should it take so long? By contrast, the thirst elicited by extracellular fluid depletion emerges rather quickly and is correlated with the decrease in extracellular fluid loss (Fitzsimons, 1961; Stricker, 1966). In addition, the elicitation of salt appetite is not correlated with either a reduction of extracellular fluid deficits or reduction of sodium (Stricker & Wolf, 1966; Wolf & Stricker, 1967). ALDO is also not necessary to demonstrate the behavioral response to extracellular fluid depletion. That is, adrenalectomized rats (but treated with a therapeutic dose of ALDO to avoid chronic sodium loss) compensate for the loss of sodium that results following body fluid depletion by ingesting both water and sodium (Wolf & Stricker, 1967). Nor is ALDO necessary for the overdrinking that occurs in sodium-depleted rats: that is, the fact that the rats drink more sodium than they need is also not dependent upon ALDO (Jalowiec & Stricker, 1970*a*, *b*), perhaps, it is due to the actions of angiotensin.

A reason for the delay in salt appetite, and perhaps the overdrinking, following a variety of natriorexigenic treatments, may have to do with an

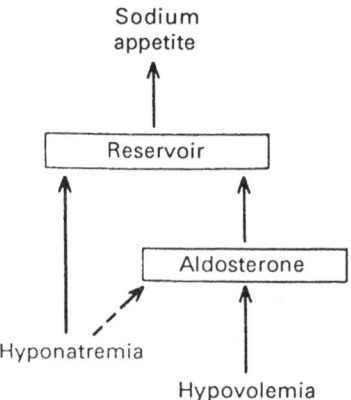

Fig. 4.10. Mechanisms of elicitation of sodium appetite according to the reservoir hypothesis (from Wolf & Stricker, 1967).

hypothetical sodium reservoir – the 'reservoir hypothesis' (Stricker & Wolf, 1966; Wolf & Stricker, 1967; Fig. 4.10). Bone is one candidate organ for this reservoir; it contains about 30% of body sodium (Bergstrom & Wallace, 1954; Forbes & McCord, 1965). Bone sodium is, in fact, reduced in body sodium-depleted rats, and can go from 152 mEq to 42 mEq of sodium per kilogram of bone (Bergstrom & Wallace, 1954; Forbes & McCoord, 1965). The mineralocorticoid hormones are known to reduce the uptake of bone sodium in bone (Streeten & Rappaport, 1963) could account for the long delay between the onset of the appetite and the natriorexigenic treatments, and for the fact that the appetite persists long after levels of sodium are not only normal but highly elevated because of the salt ingestion (Wolf & Stricker, 1967).

Other tissues rich in sodium, such as glial cells (the hepatic sites of the brain), might also act as a reservoir (Katzman, 1961). In fact, since mRNA for angiotensinogen is largely found in glial cells (Deschepper *et al.*, 1986), it may be the case that sodium depletion results in a shift in ionic transport in glial cells prompted by the genomic actions of ALDO resulting in the production of angiotensinogen which then influences neural sites to increase receptor levels (Chapter 5).

While the reservoir theory did hold promise, we know that salt appetite can be elicited rather rapidly. Recall that rats treated with high doses of renin or angiotensin elicit a salt appetite within an hour. Moreover, rats placed on a sodium-deficient diet for two days and then sodium depleted demonstrate the appetite also within an hour (Stricker, 1983). But, the excess sodium the rat usually eats may be one factor for why the appetite takes so long to develop. The animal is loaded with sodium.

It has been suggested that the surplus of sodium in the gastrointestinal tract may play an important role in the appetite (Michell, 1986). This could conceivably be the so-called sodium reservoir, particularly in ruminants. In fact, the availability of sodium within the gastrointestinal tract influences salt consumption (Michell, 1979, 1986). The accumulation of sodium in the gut can take hours to be removed, particularly after ingesting this standard diet rich in sodium (O'Kelly *et al.*, 1958); this could account for the delayed appetite. Moreover, sodium transport mechanisms within the gastrointestinal tract are affected by natriorexigenic treatments (Michell, 1979). The accelerated rate of transport in the rat placed either on a sodium-deficient, or sodium-sufficient diet, probably has something to do with the rapidity of the onset of the appetite when subsequently sodium depleted. Finally, a number of nerve endings within the gastrointestinal tract underlies Na or ionic transport, suggesting that organs within (liver) the gastrointestinal tract are acting like 'little brains' (Cooke, 1988).

SODIUM TRANSPORT

As glucose transport mechanisms have been thought to play a role in food intake (reviewed by LeMagnen, 1985), a role for sodium transport in the onset of salt appetite has been hypothesized (Woodbury, 1956; Woodbury & Koch, 1957). Blocking a hypothetical sodium pump for active sodium transport in the liver reduces salt appetite in rats (Blake, unpublished observations). This also occurs when sodium transport blockers are injected into the carotid artery of sheep (Denton *et al.*, 1969) or directly into the brain of rats (Vivas & Chiaraviglio, 1987). In rats, intragastric intubation with an agent (diphenylhydantoin) that promotes active sodium transport elevates salt intake in rats (Michell, 1979). The rapid exchange of sodium from the stomach to the hepatic portal vein may contribute to increasing the rapidity of development of the appetite. The above further suggest that a mechanism of action in salt appetite involves changes in sodium transport, promoted by the induction of protein synthesis, perhaps induced by the actions of ALDO.

POTASSIUM AND SODIUM

It has been suggested by Michell that potassium levels affect NaCl intake (Michell, 1978), and, since the turn of the century (Bunge, 1902), potassium was thought to be involved in the generation of a salt appetite. High levels of potassium were envisioned to promote sodium ingestion. The evidence for this claim is weak, or nonexistent. Infusions of potassium into the plasma do not increase sodium ingestion (Michell, 1978), and the consumption of potassium provokes elevated ALDO levels and reduced levels of renin–angiotensin (Douglas *et al.*, 1978). In fact, recall from Chapter 3 that potassium deficiency elicits sodium ingestion. And ALDO, which promotes sodium retention and potassium secretion, increases sodium consumption. Finally, loading rats with potassium does not increase sodium consumption (Wolf, 1965).

When animals are deprived of sodium, there are changes in sodium status that are reflected in plasma sodium levels (Denton, 1982). The onset of salt appetite, as indicated, in rats and sheep is not correlated with changes in plasma sodium status at the time they finally ingest the salt (Stricker & Wolf, 1969). But, rapid infusions of sodium into the systemic circulation in sodium–depleted sheep can (but does not always) inhibit salt intake (cf. Beilharz *et al.*, 1965; Weisinger *et al.*, 1983). In addition, sodium–depleted sheep can regulate their intake of sodium by bar pressing for intravenous sodium infusions (Weisinger *et al.*, 1983), and intravenous infusions of sodium can decrease the salt intake of sodium–depleted rats (Kissileff & Hoeffer, 1975). They, perhaps, have these effects by their actions directly on sodium sensors in the brain.

Osmo or sodium receptors in the brain are known to be involved in thirst (for review, see Rolls & Rolls, 1982). Sodium sensors, particularly in the herbivorous sheep, are also involved in the regulation of sodium hunger (Denton, 1982). It is well known that, not only does the brain contain sodium or osmo receptors (see Chapter 5), but angiotensin is known to interact with sodium at the level of the third ventricle; for example, infusions of angiotensin increase the water drinking that results from sodium infusions (Andersson *et al.*, 1967). Moreover, infusion of sodium into the brain results in sodium excretion (Andersson *et al.*, 1969; McKinley *et al.*, 1986*b*; see Chapter 5), which appears to be related to sodium but not to volume receptors, since isotonic infusions do not promote sodium excretion, while hypertonic NaCl does (McKinley *et al.*, 1986*b*). However, hyponatremia increases intracellular sodium brain concentration in rats (Woodbury, 1956), as does mineralocorticoid treatment (Woodbury, 1956). This may be a compensatory response to fight off the onslaught of the hyponatremia, and to preserve brain sodium.

In sodium–deficient sheep, infusions of sodium into cerebrospinal fluid reduces their salt intake (see Weisinger *et al.*, 1985*b*, 1987*a*, *b*; Denton, 1982; Fig. 4.11). Other osmotic substances that were similarly infused did not reduce their intake. Moreover, decreases in the cerebrospinal fluid of sodium in sodium–depleted sheep or cows increase their salt intake (see Weisinger *et al.*, 1985*b*; Blair-West *et al.*, 1987; Denton, 1982; Fig. 4.12), and also motivate operant behavior to obtain the sodium (Weisinger *et al.*, 1987*a*, *b*). Therefore, in herbivorous animals like the sheep and the cow, sodium sensors in the brain play a role in both the onset and satiation of the appetite. By contrast, no such relationship generally was found in rats, mice or rabbits between sodium infusions into the brain and decreases in salt intake (Denton *et al.*, 1984*a*, *b*; Epstein *et al.*, 1984; Osborne *et al.*, 1990; Tarjan *et al.*, 1984; but see Chiaraviglio & Perez Guaita, 1986). In fact, there is no correlation in the rat with the onset of the appetite and sodium concentration in the cerebrospinal fluid (Frankmann *et al.*, 1987).

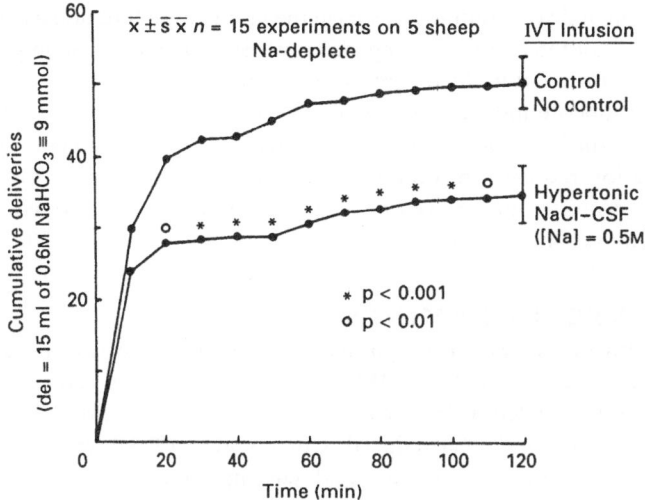

Fig. 4.11. Effect of intraventricular infusions of hypertonic sodium (0.5 M) on bar pressing for sodium in sodium hungry sheep. Infusions began 1 h before the sodium access period and for the 2 h during the test (after Denton, 1982).

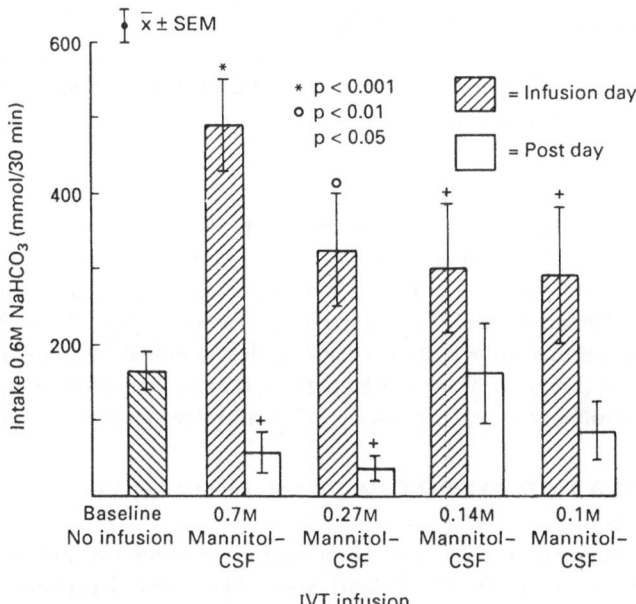

Fig. 4.12. Effect of intraventricular infusions of manitol (which decreases sodium in cerebral spinal fluid) on sodium intake of sodium hungry sheep (after Denton, 1982).

Thus, one can conclude that sodium sensors in the brain play a functional role in the regulation of salt intake of the sheep and cow. Speculatively, one might envisage that the counterpart in the rat appears to be in the liver (that is just for the satiation of the appetite and not the initiation; see Chapter 5 on satiation). In both cases, whether the brain or the liver, sodium sensors help to maintain the behavioral adaptation required in sodium homeostasis. Angiotensin, perhaps acting as a neurotransmitter, may be synthesized when the animal is deprived of sodium in both the rat and the sheep.

CENTRAL CATECHOLAMINES
The central catecholamine system is significant to the arousal of motivated behaviors (Stricker & Zigmond, 1974). While depletion of central catecholamines leaves sodium depletion-induced salt appetite intact (Stricker & Zigmond, 1974) or only somewhat reduced (Chiaraviglio, 1976a), catecholamines are known to contribute to increased salt ingestion. For example, systemic administration of adrenergic agents can increase salt ingestion (Soulairac, 1969). Noradrenalin injected into the hypothalamus increases salt intake (Chiaraviglio & Taleisnik, 1969). In addition, hypothalamic levels of noradrenaline are increased following sodium depletion (Munaro & Chiaraviglio, 1981). Finally, angiotensin-induced sodium hunger is increased with noradrenaline treatment (Chiaraviglio, 1976a). Taken together, the above results suggest that, while central catecholamines are not essential for the expression of the appetite, they do contribute to the general arousal of ingestive behavior and to sodium hunger in particular.

THYROID
Manipulations of the thyroid gland result in changes in salt intake. Hypothyroidism results in increased saline intake (Fregly & Taylor, 1964). This may result because of the loss of sodium and elevated angiotensin, and is analogous to the responses in the adrenalectomized rat (Fregly & Taylor, 1964). DOCA administration reverses this saline preference at low doses, and, at high doses, reinstates the appetite (Fregly & Waters, 1966). Thus, the salt intake following hypothyroidism is probably secondary to sodium loss and elevated angiotensin.

FOOD AND WATER DEPRIVATION INCREASE SALT INTAKE
Food deprivation or restriction has effects on sodium homeostasis and salt preference (Kaunitz et al., 1960). Food deprivation, like water deprivation, results in sodium loss in mice (Kutscher & Steilen, 1973), gerbils (Wright & Donlon, 1979) and rats (Weisinger et al., 1985a). Water deprivation results in increased salt intake (Weisinger et al., 1985a).

PATHOLOGY

The chapter closes by considering a range of pathological conditions associated with alterations of body sodium homeostasis. Older people appear less sensitive to the maintenance of extracellular fluid and sodium volume (Leaf, 1984). They should also be less responsive to natriorexigenic signals. This is not known. What is known is that pathology of the adrenal gland in a young boy resulted in alterations of salt ingestion. This is the case that was reported by Richter (Wilkins & Richter, 1940). A child was taken to Johns Hopkins Hospital by his mother because of his excessive salt ingestion; the boy ate pure salt. He was placed on a special diet, and subsequently died presumably because of the lack of salt. When an autopsy was performed on the boy it was noted that the adrenal gland had atrophied. In effect, the condition was a functional adrenalectomy, and the salt ingestion was a compensatory response.

Another condition in which salt is ingested is diabetes insipidus, a condition in which water is constantly being lost by the kidneys, resulting in excessive thirst and elevated salt intake (Titlebaum *et al.*, 1960; though also see Palmieri & Taleisnik, 1969). Angiotensin levels are elevated in this condition (Elfont & Sokol, 1987). Diabetic rats are known to express normal increases in the salt intake to adrenalectomy-induced (Palmieri & Taleisnik, 1969), and extracellular depletion-induced sodium hunger (Elfont & Sokol, 1987).

Other conditions are less associated with salt ingestion, but are associated with alterations of body fluid or sodium balance. For example, psychotic syndromes have been associated with alterations of water (Berl, 1988) and with low sodium balance (Burnell & Foster, 1972). In addition, too much sodium can lead to seizures (Glaser, 1964), and can trigger migraine headaches (Brainard, 1976).

Sodium deprivation during development reduces growth and delays the development of sexual maturity in rats (Mouw *et al.*, 1978). It also reduces the size of the litter (Bursey & Watson, 1983). Moreover, sodium levels of the mother can affect the sex ratio in rat litters; a high sodium diet during pregnancy results in a lower ratio of males being born (Bird & Contreras, 1986). Recall that extracellular fluid deprivation during pregnancy in rats elevates the sodium intake of their offspring when they are mature; it has been suggested that excessive vomiting and reduction of extracellular fluids during pregnancy may result in increased salt intake in the offspring (Nicolaidis *et al.*, 1990).

Most of the hormones that have been discussed can also produce pathological conditions. The renin–angiotensin cascade has long been known to be involved in several forms of hypertension (Shiono & Sokabe, 1976; Horky & Gregorova, 1980). The bulk of the evidence suggests that angiotensin affects cardiovascular and brain function. The solitary nucleus in particular, appears to be involved in some of the cardiovascular effects that result from angiotensin (Ferrario *et al.*,

1972). In fact, increased binding of angiotensin II in the solitary nucleus has been reported in hypertensive rats (Plunkett & Saavedra, 1985), and injections of angiotensin into the solitary nucleus promotes changes in blood pressure (Casto & Phillips, 1985). The exaggerated salt appetite of some hypertensive strains is decreased by blocking angiotensin synthesis in the brain (DiNicolantonio *et al.*, 1982).

Mineralocorticoids provoke changes in blood pressure; intracerebral infusions of ALDO can increase blood pressure (Gomez-Sanchez, 1986). DOCA-salt induced hypertension increases catecholamine synthesis in both the hypothalamus and in the solitary nucleus (Hwang *et al.*, 1984); medial hypothalamic knife cuts decrease DOCA-salt induced hypertension (Bealer, 1984). In addition, mineralocorticoids are known to affect cardiovascular regions of the brain, that is, the solitary nucleus (Spyer, 1981), and lesions within the intermediate region of the solitary nucleus are known to promote cardiovascular disturbances (Reis, 1981). The solitary nucleus, in particular the intermediate region, is involved in cardiovascular, blood pressure and body fluid regulation (Spyer, 1981; Reis, 1981).

Various pathological conditions may be associated with ANF. For example, congestive heart failure may result from damage to ANF-containing endocrine cells within the heart. Strains of hamsters, for example, known to develop congestive heart failure, are deficient in ANF (Chimoskey *et al.*, 1984; Sonnenberg *et al.*, 1983).

One can clearly see that the hormones that regulate sodium homeostasis also have profound affects on the brain and on cardiovascular function. But, stress also affects cardiovascular functions, perhaps by promoting ACTH secretion and salt intake (Denton *et al.*, 1984a; Chapter 2). ACTH infusions can result in changes in blood pressure and hypertension (Butkus *et al.*, 1985).

Genetic factors do contribute to the expression of hypertension. For example, some strains of rats which, by their genetic predisposition ingest salt develop hypertension; those without this predisposition, that ingest salt, do not (e.g. Dahl strain, Dahl *et al.*, 1962). Within genetic strains of rats, high blood pressure can be dissassociated from salt intake (Yongue & Myers, 1988). Genetic factors figure in whether rats with hypertension will avoid or ingest salt. That is, a variety of forms of hypertension in rats can actually promote the avoidance of salt (Wolf *et al.*, 1965). This is clearly an adaptative response. However, other genetic strains of hypertensive rats actually prefer the salt (Catalanotto *et al.*, 1972; Bertino & Beauchamp, 1988).

But, what about the ingestion of sodium? Does it result in hypertension? High sodium intake has long been thought to be an etiological factor in hypertension (e.g. Dahl, 1958; Dahl & Love, 1954). Infants fed high sodium diets were thought to be particularly susceptible to hypertension (Dahl *et al.*, 1963; Mayer, 1969). But, it still remains unclear, at this date, whether sodium causes hypertension (e.g. Kurtz & Morris, 1985), whether humans with hypertension have an enhanced salt appetite, or whether gustatory mechanisms are altered in hypertensive patients (Mattes, 1984).

SUMMARY

In this chapter the principal hormones of sodium homeostasis have been placed in a physiological (organ-centered) context in which they were depicted as maintaining sodium and extracellular fluid balance. It is thus clear that many organs of the body participate in the regulation of sodium, and, perhaps, molecular mechanisms that involve sodium transport-dependent protein synthesis are fundamental to many of the regulatory controls.

5 Neural circuits underlying salt intake

I am the E Power Biggs of Electron Microscopy

I am the Dali of Luxal Blue
I have captured the septal confluence
In a radiance of sky blue blades of grass
And bright pink perikarya
I have sketched the sagittal arch of the fornix
With blue–black hematoxylin
I have sculptured neurons with black mercury
As Golgi did before me
And I have filled fasciculi with droplets of silver
Like pebbles among the astrocytes
Strewn on the ventralis of the brown earth

In short I have loved the brain – its oligodendroglia
I have stroked the seventh nerve from flocculus to terminus
Till the genu arched and purred
And puffed my cheeks to nucleus globosus et rotundus
Feeling fat and full
Yet I tore penduncles from the cerebellum
Like a lover tears petals from a flower
Or a child limbs from an insect
And I crossed degenerated fragments of the hippocampal commissure
Where Pontius Pilate stole the gyrus from the hemisphere
To find it was not me
Gazing across the anterior horns of the universe

GEORGE WOLF

INTRODUCTORY

In this chapter, an anatomical circuit, that underlies the behavioral regulation of
body sodium homeostasis, is outlined. Many of the sites are involved in other

regulatory behaviors (e.g. thirst and hunger) and subserve alimentary functions, in general, in addition to reproductive functions. The anatomical circuit outlined, therefore, has implications more general than that of sodium hunger *per se*. The circuit is still somewhat speculative, but warranted given our current knowledge.

The chapter is organized around hormonal receptor localization, sodium sensors and their functions, and anatomical connectivity of these areas. Lesion studies will help us construct functional interpretations of the anatomy. Some of the main points are: 1) The anatomical circuit includes angiotensin and sodium-sensitive sites in the anterior third ventricular region. This region is important for thirst and sodium hunger, extracellular fluid and cardiovascular regulation. 2) Also important are mineralocorticoid-sensitive sites in the medial amygdala, and perhaps the medial preoptic and bed nucleus of the stria terminalis. The salt intake to the mineralocorticoids, and the neural circuit that underlies it, may be tied to the demands for sodium during reproduction. Finally, 3) The central nucleus of the amygdala is generally involved in taste-motivated behaviors, and is also known to be involved in cardiovascular control.

THE BRAIN ANGIOTENSIN SYSTEM

The localization of brain angiotensin II receptor systems, involved in thirst and sodium hunger, is being studied by *in vivo* and *in vitro* autoradiography and mRNA hybridization techniques. Recall that angiotensin of renal origin does not cross the blood–brain barrier. *In vivo* studies have shown that blood-borne angiotensin reaches only those structures in the brain that are outside the blood–brain barrier, such as the circumventricular organs (CVOs e.g. subfornical organ, area postrema, organum vasculosum of the lamina terminalis, OVLT, pineal gland), and the pituitary gland (for review, see Lind, 1988). The CVOs provide sites for the brain to obtain information about both peripheral blood and CSF levels of many solutes.

In vitro studies of the brain (e.g. McKinley *et al.*, 1986*a*), using pharmacological and radioautographic techniques have localized angiotensin receptors in: 1) CVOs that can be reached by blood-borne angiotensin, such as the subfornical organ, the area postrema, pituitary gland, pineal gland and the OVLT; 2) hypothalamic regions, such as magnocellular parts of the paraventricular nucleus, supraoptic nucleus, nucleus medianus and the lateral hypothalamus; 3) the zona incerta; 4) septum and thalamus; 5) corpus striatum; 6) the medial and central nucleus of the amygdala; 7) olfactory areas; 8) parabrachial nucleus; and 9) the nucleus of the solitary tract. With few exceptions, all of these are sites within a potential neural circuit for hydromineral balance.

Angiotensin-containing neurons and fiber projections have also been localized using immunohistochemical techniques (Lind *et al.*, 1985; Lind, 1988),

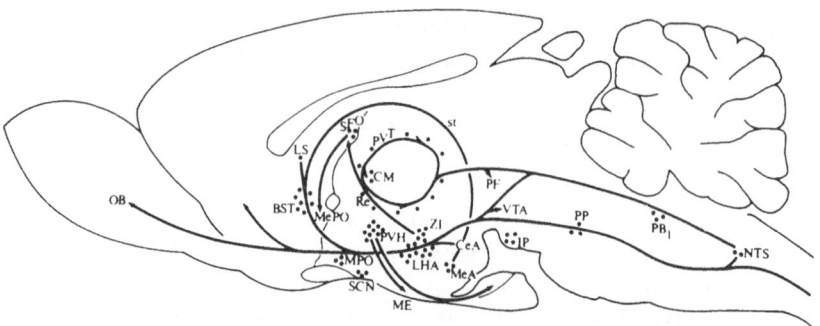

Fig. 5.1. This schematic mid-sagittal view of the rat brain illustrates the major AII–immunoreactive cell groups and fiber system described in the present study. Arrowheads are drawn to indicate the orientation of projections in cases where this is obvious (from Lind, 1988). See text below for relevant structures.

and angiotensin–containing neurons and terminal fields have been found in: 1) the subfornical organ; 2) nucleus medianus; 3) perifornical lateral hypo-thalamic region; 4) rostral parts of the zona incerta; 5) dorsal region of the bed nucleus of the stria terminalis; 6) preoptic region; 7) medial and central nuclei of the amygdala; 8) lateral parabrachial nucleus; and 9) the nucleus of the solitary tract at the edges of the area postrema (Lind *et al.*, 1985; McKinley *et al.*, 1986*a*). This is schematically depicted in Fig. 5.1.

Renin, the rate-limiting enzyme for angiotensin synthesis, has also been found in several regions of the brain using immunohistochemical techniques. The highest amounts are found in the hypothalamus and amygdala (Ganong, 1984). Many of these brain regions are either involved in the regulation of body fluids or in cardiovascular control (e.g. Johnson, 1985; Reis, 1981).

In addition, angiotensinogen mRNA production has been localized in the brain, using hybridization methods (Lynch *et al.*, 1987; Angulo *et al.*, 1988). Angiotensinogen mRNA appears to be localized largely to astrocytes (Chapter 4). Angiotensinogen mRNA has been found in many of the areas where there are angiotensin receptors and angiotensinergic fiber pathways. Finally Ang II receptors in cultured neuron-like cells apear to act through guanine nucleotide binding proteins (Reagan *et al.*, 1990)

It is clear then that all of the components of a renin–angiotensin system are in the brain and are separate, anatomically, from the renal renin–angiotensin system. Thus, the idea that the hormone in the brain, perhaps, acts as a neurotransmitter in the genesis of sodium hunger or thirst is possible (Denton, 1982; Chapter 4).

ALDOSTERONE

There is no blood–brain barrier for ALDO. Furthermore, the brain does not produce, or synthesize, ALDO. The only source is from the adrenal gland.

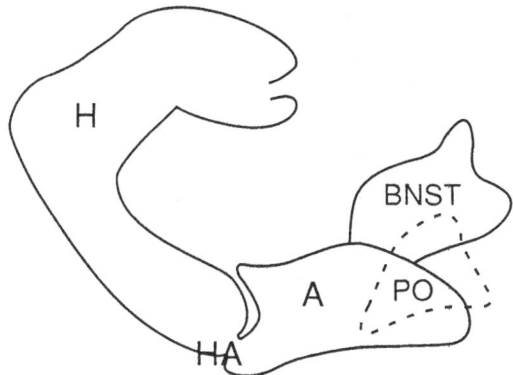

Fig. 5.2. Depicts brain regions known to concentrate aldosterone (H = hippocampus; A = amygdala; HA = amygdala–hippocampal transition zone; PO = preoptic area; BNST = bed nucleus of the stria terminalis) (from Schulkin, unpublished observations).

Central ALDO receptor regions have been uncovered using autoradiography, cell fractionation and receptor competition studies, together with cytosolic receptor analysis (e.g. Birmingham *et al.*, 1986; Coirini *et al.*, 1985). While both glucocorticoids and mineralocorticoids can compete for the same neural receptors (Chapters 2 and 4), the receptors for the two hormones can also be distinguished in the brain, e.g. by blocking all glucocorticoid sites with a specific antiglucocorticoid, and thereby leaving unoccupied the mineralocorticoid receptor sites, or by saturating the brain with corticosterone and again leaving open specific mineralocorticoid binding sites (e.g. Coirini *et al.*, 1983). This has resulted in the localization of the mineralocorticoid receptors to specific brain regions. They include: 1) hippocampus; 2) cortico-medial or central amygdala; 3) subiculum; 4) medial–preoptic hypothalamus; 5) bed nucleus of the stria terminalis. These regions are schematically depicted in Fig. 5.2.

The mineralocorticoid receptor has been cloned (Arriza *et al.*, 1988; Evans & Arriza, 1989). Using *in situ* hybridization histochemistry, the distribution of the mRNA has been outlined in the brain (Arriza *et al.*, 1988; Chao *et al.*, 1989). The distribution is extensive and overlaps with many of the same regions as the mineralocorticoid receptor sites (Arriza *et al.*, 1988). It has been suggested that a common molecular genetic code is the basis for a whole family of hormones (Evans, 1988).

Earlier studies tended to emphasize the hippocampus as a major site for aldosterone (e.g. Stumpf & Sar, 1979). But the hippocampus may not be

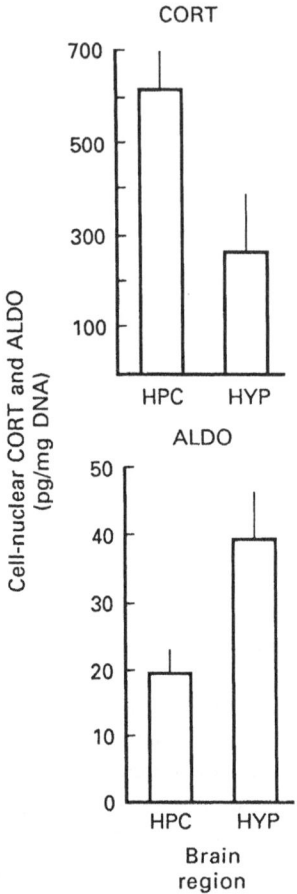

Fig. 5.3. Brain cell-nuclear CORT and ALDO. (HPC, hippocampus; HYP, hypothalamus (after Yongue & Roy, 1987).

importantly involved in sodium hunger, because lesion studies show that the region can be removed with no effect on sodium hunger (Murphy & Brown, 1970). On the other hand, the preoptic–hypothalamus and its connections to the amygdala and bed nucleus do seem to be involved in sodium hunger. Thus Yongue & Roy (1987) found preferential binding for ALDO in preoptic–hypothalamic tissue over hippocampal tissue (Fig. 5.3) in adrenally intact rats using subcellular fractionation and radioimmunoassay of ALDO. These results suggest that ALDO binds preferentially to the preoptic–hypothalamic region. Furthermore, while the hippocampus is well known to be involved in spatial performance (e.g. Olton & Markowska, 1989),

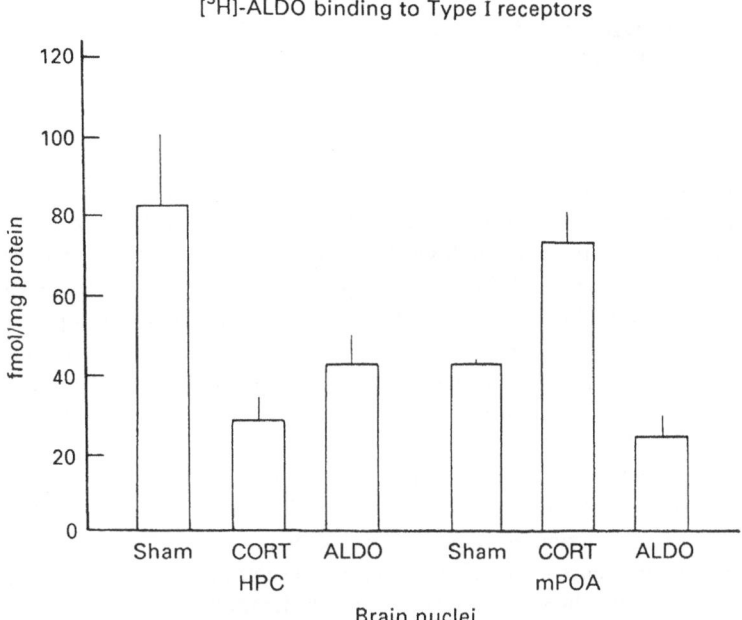

Fig. 5.4. Aldosterone binding in rats treated with (CORT, ALDO, or vehicle, HPC, hippocampus; mPOA, Medial preoptic area) (from Coirini, Schulkin & McEwen, 1988).

the preoptic–hypothalamic region is known to be involved in body fluid homeostasis (e.g. Fitzsimons, 1979).

Further research has corroborated and extended this observation. CORT preferentially binds to the hippocampus and ALDO preferentially binds to the preoptic–hypothalamic region (Brinton & McEwen, 1987). As noted in earlier chapters, CORT can actually increase the level of ALDO binding in preoptic–hypothalamic tissue (Brinton & McEwen, 1987; Coirini *et al.*, 1988; see also Chapter 2 and 4). In this study, rats were adrenalectomized and treated with CORT and ALDO for three days before being sacrificed.

Rats treated with CORT showed increased cytosolic binding of ALDO in the preoptic–hypothalamic region, but not in the hippocampus (Fig. 5.4). Since there are a greater number of CORT than ALDO sites in the brain (Chao *et al.*, 1989), one result of elevated glucocorticoid levels during sodium depletion may be the increase of ALDO binding in selective sites of the brain, and therefore an increase in the natriorexigenic potency of ALDO (Chapter 2 and 4). Thus, the fact that ALDO-induced sodium hunger is augmented with CORT treatment is not surprising.

SEPARATE BRAIN REGIONS FOR THE NATRIOREXIGENIC HORMONES

There are separate neuroanatomical systems in the brain for different dipsogenic signals (for review, see Rolls & Rolls, 1982). Physiological and behavioral studies indicate that there are separate mechanisms in the brain for the salt appetite that results from angiotensin II or ALDO. Either hormone when given alone can stimulate the appetite (Chapter 2). Recall that mineralocorticoid treatment decreases renin levels in doses that stimulate a salt appetite (Pettinger *et al.*, 1971); in the absence of ALDO (adrenalectomy), angiotensin levels are four to six times higher in these sodium-hungry rats (Sakai & Epstein, unpublished observations). In addition, captopril, an inhibitor of angiotensin synthesis, when given centrally, reduces the salt appetite that results from sodium deficiency but does not reduce mineralocorticoid-induced salt appetite (Weiss *et al.*, 1986).

But, when both hormones are blocked centrally, the salt appetite that results from sodium deficiency is completely suppressed (Sakai *et al.*, 1986). Moreover, if mineralocorticoid action is removed via adrenalectomy and angiotensin biosynthesis is prevented, or if angiotensin receptors are blocked with competitive analogs, then the appetite for salt is abolished in the adrenalectomized rat (Chapter 4). By contrast, ALDO-induced salt appetite, but not depletion-induced salt appetite, is abolished by specific mineralocorticoid blockers (Wolf, 1969a). Thus, one can see that either hormone working independently of the other can stimulate the appetite. Of course, the natural state is for both hormones to be elevated during sodium deficiency and to act in unison.

SODIUM HUNGER IS DEPENDENT UPON THE FOREBRAIN

The level at which a behavior is organized by the brain can be instructive with regard to its evolutionary emergence (Jackson, 1884). As mentioned in Chapter 3, forebrain structures are essential for both thirst and salt appetite. Chronic decerebrated rats (Fig. 5.5), unlike intact rats, do not increase their ingestive responsiveness to orally infused NaCl (actual intake and taste reactivity) when rendered sodium hungry by sodium depletion as they do for sucrose when rendered hungry (Grill, 1980; Grill *et al.*, 1986). One set of results is shown in Fig. 5.6. They also do not increase their water intake to dipsogenic stimuli (Grill & Miselis, 1979). Their excretion of sodium is mostly normal. While decerebrate rats are known to demonstrate comparable taste reactivity to different tastes when taste solutions are infused into the oral cavity (Grill & Norgren, 1978c), they do not respond the same way to NaCl (Flynn & Grill, 1988; Chapter 3). Thus in contrast with the behavioral and autonomic responses

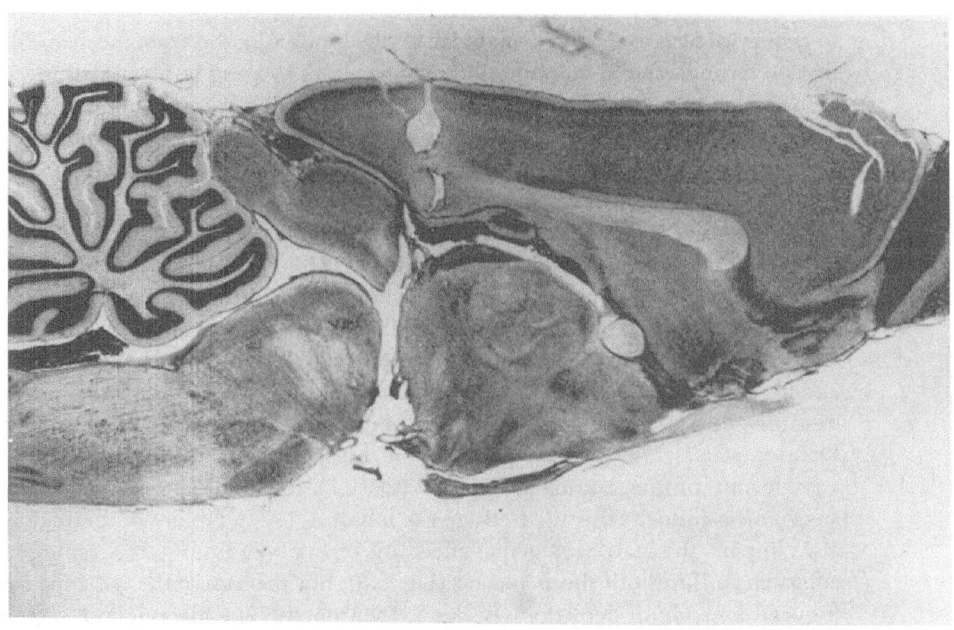

Fig. 5.5. Representative sagittal section of the brain of a supracollicular chronic decerebrate rat (from Grill, Schulkin & Flynn, 1986).

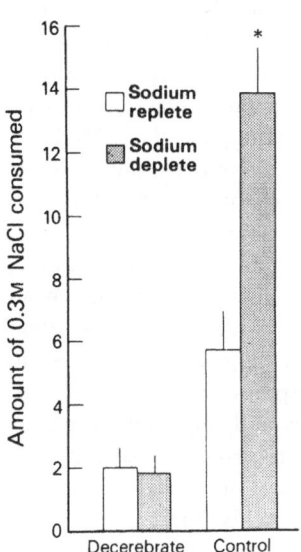

Fig. 5.6. Mean intraoral 0.3M NaCl ($\pm$ SEM) of decerebrate and control rats while sodium replete and following the natriorexigenic treatments (sodium deplete) (from Grill, Schulkin & Flynn, 1986).

to energy deficits that can be mediated by the hindbrain, the brain mechanisms that govern water and salt ingestion may not have evolved until animals more complex than fish and amphibia emerged from the sea (Schulkin, 1986). Consequently, sodium hunger is dependent on the forebrain for its expression.

VENTRAL LAMINA TERMINALIS AND ANGIOTENSIN-INDUCED SODIUM HUNGER

Consider, now, regions of the brain that may be importantly involved in angiotensin-induced sodium hunger. Midline structures surrounding the third ventricle are known to be importantly involved in angiotensin-induced thirst and hemodynamical hydromineral balance (e.g. Buggy & Johnson, 1977). For example, lesions of the anterior third ventricular region (AV3V, Fig. 5.7; DeLuca et al., 1990) which damage the nucleus medianus, and organum vasculosum of the lamina terminalis (OVLT) impairs centrally delivered angiotensin-induced thirst (e.g. Buggy & Johnson, 1977). Lesions of this region also impair the natriorexigenic effect of centrally administered renin or angiotensin II in both sheep and rat (Fig. 5.8); but they leave the salt appetite induced by sodium depletion (Bealer & Johnson, 1979; Chiaraviglio & Perez Guaita, 1984; Denton et al., 1986), and mineralocorticoid-induced sodium hunger relatively intact (DeLuca Jr et al., 1990; Fitts et al., 1990).

More generally, lesions within the AV3V region also result in chronic hypernatremia and reninemia (Buggy & Johnson, 1977). They also reduce natriuresis (Bealer & Johnson, 1979). Recall that this brain region is a major target site for cells in the cardiovascular parts of the parabrachial region (Saper, 1983), and also responds to hemodynamic changes (Buggy et al., 1984). Recall that the anterior third ventricular region of the brain contains sodium sensors that are involved in osmotic regulation, thirst, and activation of ADH secretion (Chapter 4). Excretion of NaCl, following infusions of hypertonic NaCl, is impaired following lesions of the AV3V in both rats and sheep (McKinley et al., 1986b), and perhaps it is the region in the sheep where the initiation and satiation of the appetite is regulated by sodium levels (Denton, 1982). Therefore, the AV3V is ideally suited to monitor the changes in body fluid dynamics and to help generate the behavioral response. Also, recall that this brain region has angiotensin-containing neurons and angiotensinergic terminals from brain regions known to be involved in cardiovascular regulation and baroreceptor reflexes; it is richly connected to brainstem regions known to be involved in similar functions (Lind, 1988). It is richly interconnected with various other brain regions known to be involved in body fluid homeostasis, e.g. subfornical organ, medial preoptic region, parabrachial region (Camacho & Phillips, 1981; Saper et al., 1983).

There is also direct evidence from central infusion studies that the ventral

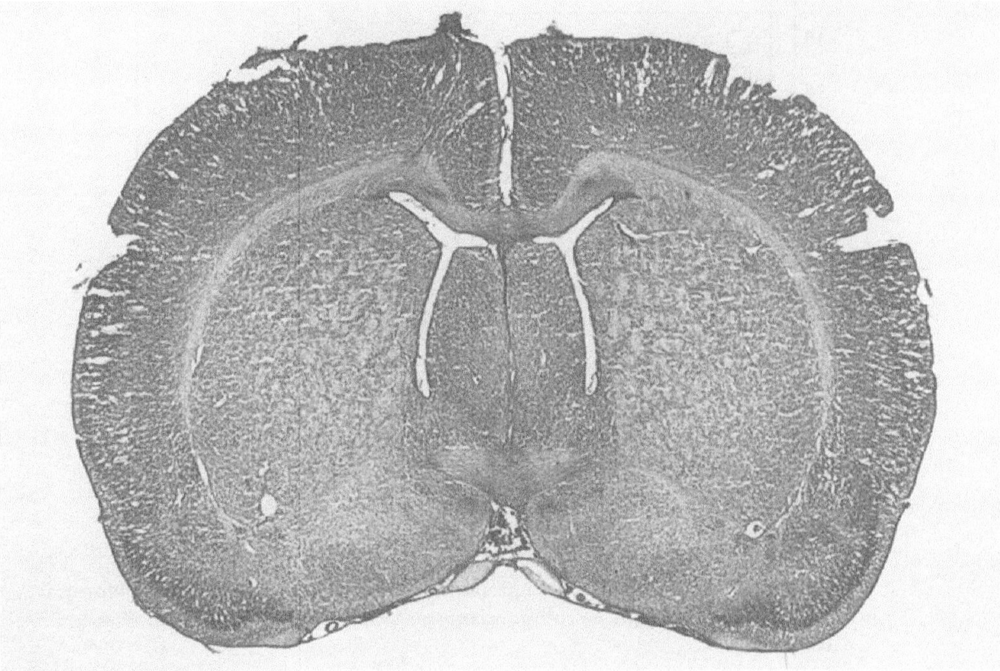

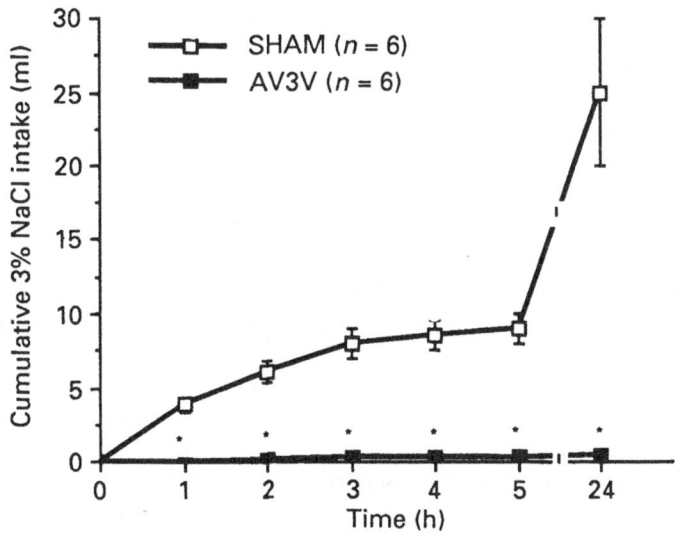

Fig. 5.7. Photomicrograph of AV3V lesion (from DeLuca *et al.*, 1990).

Fig. 5.8. Increased ingestion of sodium to intracerebroventricular injections of 100 ng renin (DeLuca *et al.*, 1990).

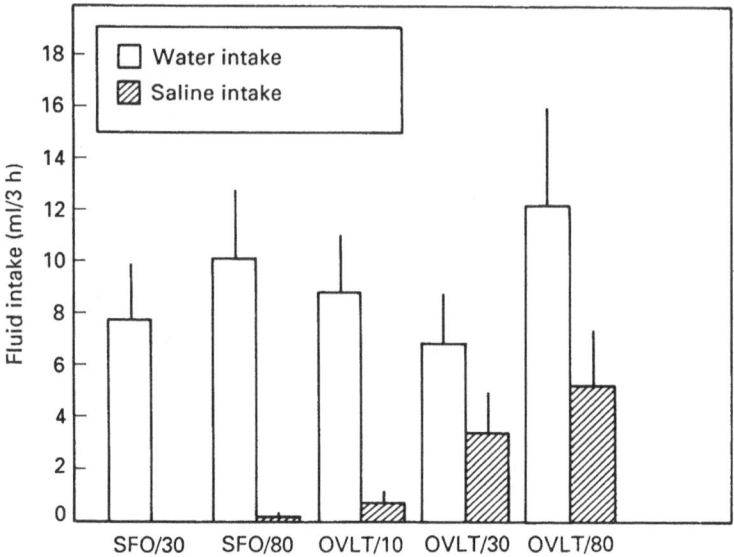

Fig. 5.9. Fluid intakes by rats having infusions of angiotensin II near the subfornical organ (SFO) or organum vasculosum laminae terminalis (OVLT), (from Fitts & Mason, 1990).

nucleus medianus has an important role in angiotensin-induced sodium hunger. Intracerebral injections of angiotensin into the region of the nucleus medianus or OVLT result in a robust salt appetite (Fitts & Mason, 1990; Fig. 5.9); by contrast, infusions into the subfornical organ, known for its involvement in thirst generated from renally derived renin–angiotensin (e.g. Simpson, 1975), do not elicit this response, nor do infusions into the region surrounding the area postrema in the lower brainstem elicit this increase in salt ingestion. These facts, along with the above, suggest that the nucleus medianus or OVLT may be important sites for angiotensin-induced sodium hunger. The fact that the major sites for angiotensin receptors and angiotensinogen mRNA are in this region is more evidence for this claim.

   In summary, the evidence suggests that the AV3V region of the brain is important to the angiotensinergic contribution in the arousal of sodium hunger or thirst to body sodium depletion, or extracellular fluid loss. These sites also contain sodium or osmo-receptors (Thrasher *et al.*, 1982). Perhaps, the sodium hunger, generated from angiotensin's action, is tied to hemodynamic and hydromineral balance.

MEDIAL AMYGDALA INVOLVEMENT IN MINERALOCORTICOID-INDUCED SODIUM HUNGER AND THE FEMALE'S ENHANCED AVIDITY FOR SODIUM

As the AV3V region is responsive to angiotensin and the regulation of extracellular water and sodium balance, perhaps other brain regions are important for the mineralocorticoid natriorexigenic actions. One likely site is the amygdala. C. Judson Herrick (1948) had described the amygdala as a region where the old and the new meet; it is old cortex but new brain, is involved in regulatory functions like the hypothalamus, and is ideally suited to play a role in some aspect of sodium hunger. The amygdala is well known for its role in feeding, taste aversion learning and sexual behavior (Braun *et al.*, 1982).

Damage to the medial amygdala has been reported to impair DOCA-induced salt appetite (Nachman & Ashe, 1974). Reversible chemical damage of the amygdala impairs sodium depletion-induced salt appetite (Zolovick *et al.*, 1980). Also note that complete ablation of the amygdala abolishes depletion-induced sodium hunger (Cox *et al.*, 1978). Recall that the medial region of the amygdala contains mineralocortoid receptors. Importantly, electrolytic damage to the medial amygdala (Fig. 5.10), specifically the medial nucleus of the amygdala, abolishes or reduces the appetite for salt that results from mineralocorticoid treatment (Schulkin *et al.*, 1989*b*; Nitabach *et al.*, 1989; Fig. 5.11). That is, rats given different doses of ALDO coupled with a sodium-deficient diet over a two-day period, once weekly, in a counterbalanced design, do not ingest the sodium if they have medial amygdala damage. Similarly, rats injected with pharmacological doses of DOCA, while maintained on the normal sodium diet, also do not ingest the sodium with medial amygydala damage. Lesions outside this region (caudal or unilateral), but within the amygdala, did not produce this effect (Nitabach *et al.*, 1989). By contrast, rats with medial amygdala damage were responsive to sodium depletion-induced (Fig. 5.12) or adrenalectomy-induced salt appetite (Schulkin *et al.*, 1989*b*). Recall that angiotensin, in this case alone, can presumably act on other brain structures to generate the appetite for salt (e.g. nucleus medianus, and see section on the central nucleus of the amygdala). These results are consistent with the idea that there are separate neuroanatomical systems in the brain for the natriorexigenic effect of each hormone.

The medial region of the amygdala is not only a prime target of mineralocorticoids, but also of other steroid hormones (McEwen, 1989). It also receives primary olfactory projections from the olfactory bulb and the vomeronasal organ involved in reproductive behavior (Scalia & Winans, 1975; Kevetter & Winans, 1981*a*, *b*). The medial nucleus receives not only primary olfactory projections but also visceral projections from the solitary nucleus and

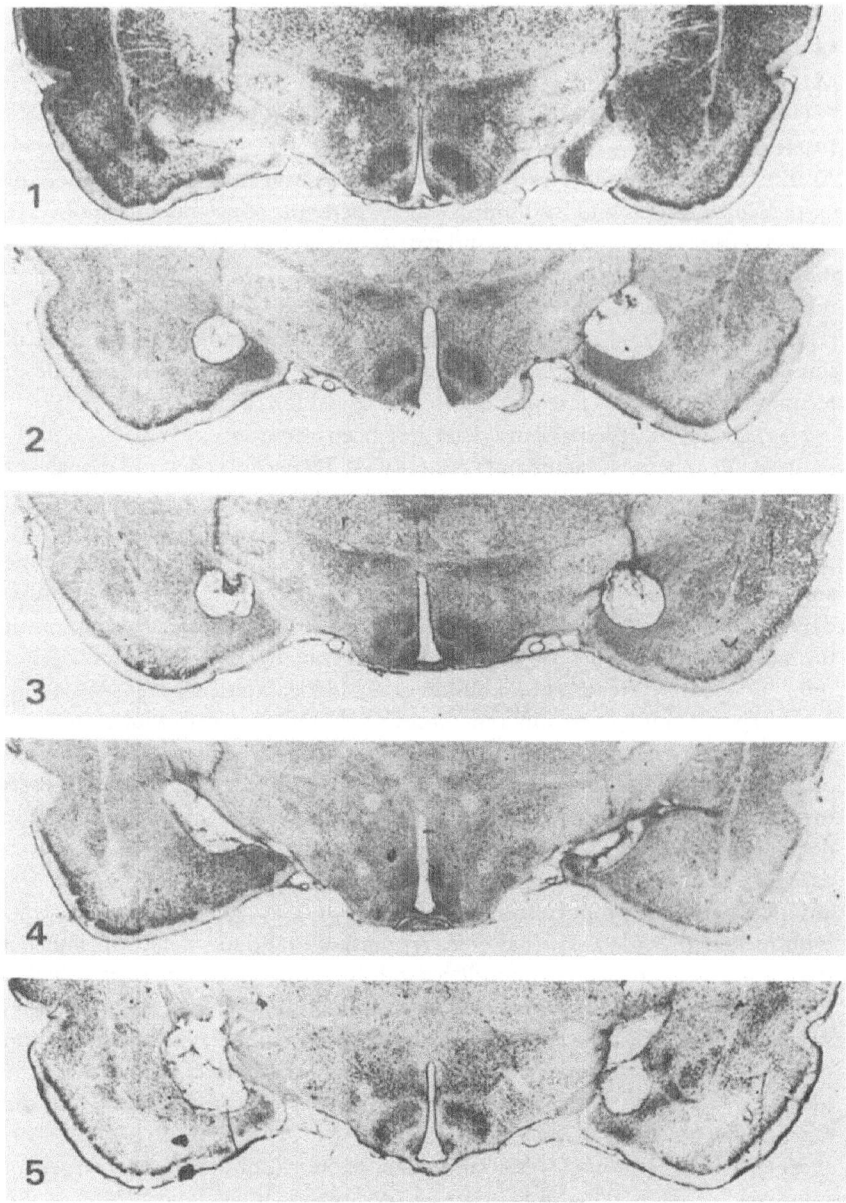

Fig. 5.10. Representative photomicrographs of five individual rats with bilateral amygdala damage. (In each case, there is significant damage to the medial region, specifically the medial nucleus, from Schulkin, Marini & Epstein, 1989.)

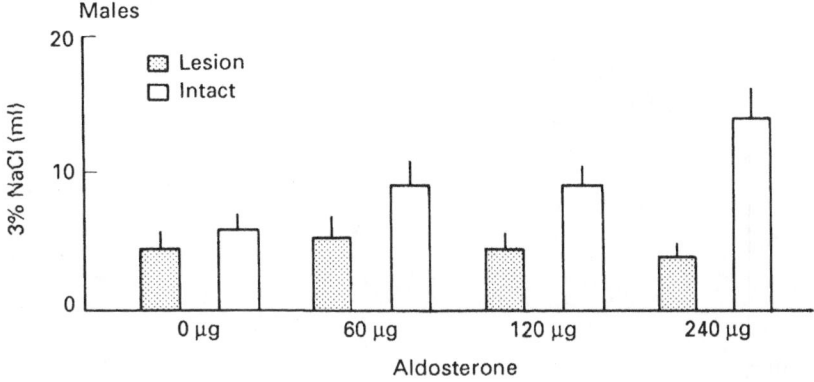

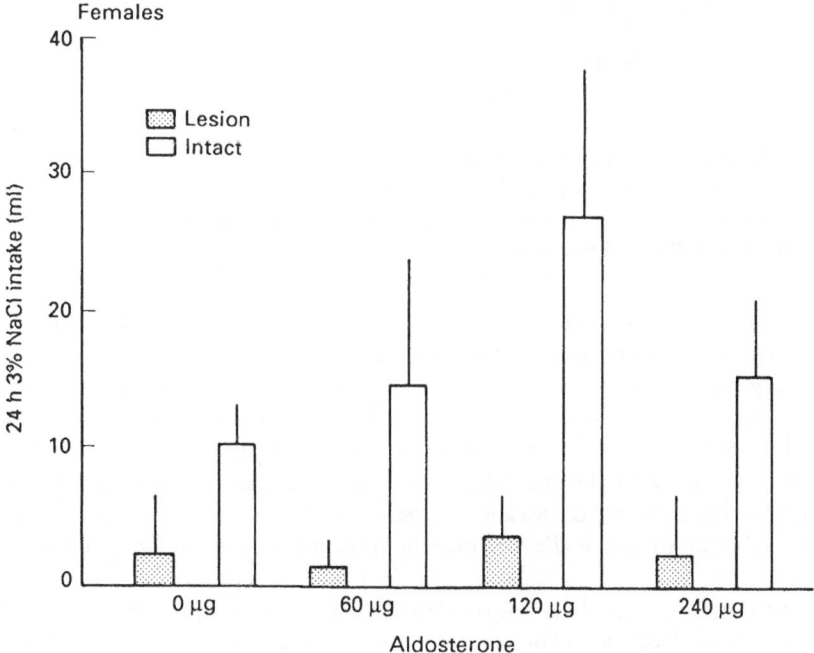

Fig. 5.11. 24-h ingestion of salt following aldosterone treatment in animals with (lesion) and without (sham lesion) damage to the medial amygdala (from Marini, Schulkin & Epstein, 1986; Nitabach, Schulkin & Epstein, 1989).

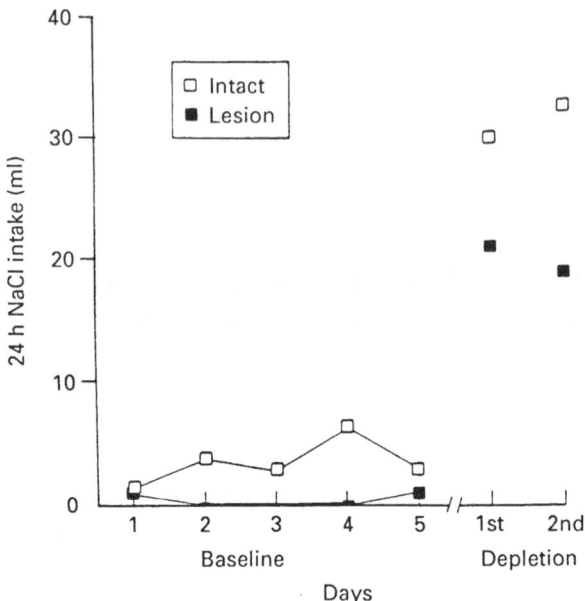

Fig. 5.12. 24-h NaCl intakes in lesion and intact rats; intakes shown are both need-free (prior to any sodium-depleted treatment over a 5-day period) and need-induced (after the rats' first and second episodes of sodium depletion (from Nitabach, Schulkin & Epstein, 1989).

the parabrachial region of the brainstem (e.g. Deolmos *et al.*, 1985). In addition, this region receives projections from regions of the forebrain which play a role in body fluid homeostasis (Deolmos *et al.*, 1985). It is also known, as I have indicated, to contain both angiotensin receptors and angiotensinogen mRNA in the medial as well as the central nucleus. The brain region is richly interconnected with the medial portions of the preoptic region (Krettek & Price, 1978; Simerly & Swanson, 1986). Perhaps, most important is its connectivity with the medial bed nucleus of the stria terminalis (Deolmos *et al.*, 1985).

The medial amygdala is not only richly connected with the preoptic, bed nucleus, hypothalamic and hindbrain regions (Nauta, 1961; Cowan *et al.*, 1965), but also richly projects to a number of neocortical sites (see Deolmos *et al.*, 1985), including the frontal cortex. In fact, most of the sensory systems have projections to nuclei of the amygdala which then act to integrate as it passes the information to the neocortical areas (Turner *et al.*, 1980). It is, therefore, ideally suited to integrate information from a variety of sources in the maintenance of sodium homeostasis.

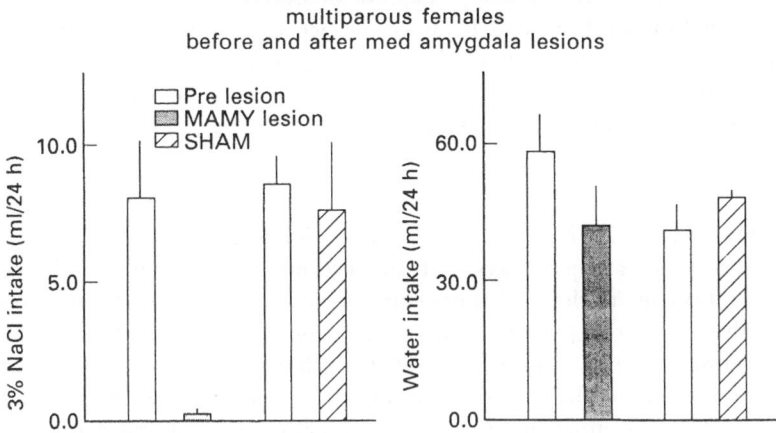

Fig. 5.13. Salt and water intake of multiparous females before and after medial amygdala damage (from DeLuca, Jr & Schulkin, unpublished observations).

Consider the hypothesis that the mineralocorticoids are acting on a steroid-sensitive circuit that underlies other steroid-induced behaviors. The medial nucleus of the amygdala is known to be involved in other steroid-induced behaviors. For example, the medial nucleus of the amygdala, as well as the medial preoptic nucleus and bed nucleus of the stria terminalis, are involved in steroid-induced male copulatory behavior (e.g. Lehman & Winans, 1980). The gonadal steroids bind to these three sexually dimorphic regions which are differentiated by the actions of the gonadal steroids during critical stages in development (e.g. Goy & McEwen, 1977). The fact that all three regions bind the mineralocorticoid hormones suggests that these three forebrain sites may constitute a steroid circuit in the brain underlying regulatory and reproductive behaviors. These brain regions themselves are projected to by many of the same brainstem and forebrain sites involved in body fluid homeostasis, and, in many instances, project back to these same sites in the brainstem (Deolmos *et al.*, 1985).

This same steroid circuit may therefore underlie the female's ingestion of salt (Chapter 2), and the increases of salt intake following natriorexigenic treatments resulting from mineralocorticoid treatment. With regard to the female's ingestion of salt, recall that: pregnant females ingest large amounts of salt and that ALDO levels are higher during pregnancy. Damage to the sexually dimorphic medial nucleus of the amgydala decreases the salt intake of female rats (Nitabach *et al.*, 1989; De Luca Jr *et al.*, unpublished observations; Fig. 5.13). The interactions between the medial nucleus of the amygdala and the bed nucleus of the stria terminalis and perhaps the preoptic region may be the basis for the sexual dimorphism of salt intake. In fact, the postulated aldosterone circuit in the brain, which includes these three regions, may be the anatomical and neuroendocrine basis of the elevated salt intake in the female and of the elevated salt intake during pregnancy and lactation.

DISCONNECTION OF ANG AND ALDO CIRCUIT

Recall that there are pathways that connect midline structures with the amygdala. Damage to the medial or central region of the amygdala interferes with pathways to the preoptic and bed nucleus of the stria terminalis by disconnecting the stria terminalis. Thus, the effects of lesions of medial or central amgydala (see below) may be due to the interruption of the fiber pathway. But, disconnecting the stria terminalis does not interfere with mineralocorticoid-induced sodium hunger, or depletion-induced sodium hunger (Black *et al.*, unpublished observations). However, it is known that disconnection of both fiber pathways from the amygdala to midline structures of the anterior forebrain disrupts sodium depletion-induced sodium hunger (Chiaraviglio, 1971). Specifically, electrolytic lesions of the stria terminalis and the amygdalafugal pathways together (but not separately) severely reduce the salt ingestion that results from sodium depletion in rats. These results suggest that disconnecting mineralocorticoid and angiotensin sites from one another interferes with their synergistic action in the genesis of sodium hunger in the rat.

SUMMARY STATEMENT OF THE TWO INDEPENDENT SYSTEMS FOR THE TWO NATRIOREXIGENIC HORMONES

The above results, when taken together with the results of our amygdala lesions, lead us to the hypothesis that there are separate neural systems for angiotensin- and mineralocorticoid-induced salt appetite. By separate systems, I mean that the central mode of action of the two hormones is on different neural circuits, one for angiotensin and the other for ALDO. As a tentative hypothesis, angiotensin appears to have its effects within the medial structures of the anterior third ventricular region, while ALDO probably acts within the medial region of the amygdala, specifically within the medial nucleus of the amygdala and possibly within the bed nucleus of the stria terminalis via its reciprocal connections with this region (but by some other route than the stria terminalis pathway, perhaps the amygdalafugal pathway). But, there must be a region in the brain where the effects of the two hormones interact in the genesis of the synergistic response to simultaneously elevated levels of these hormones.

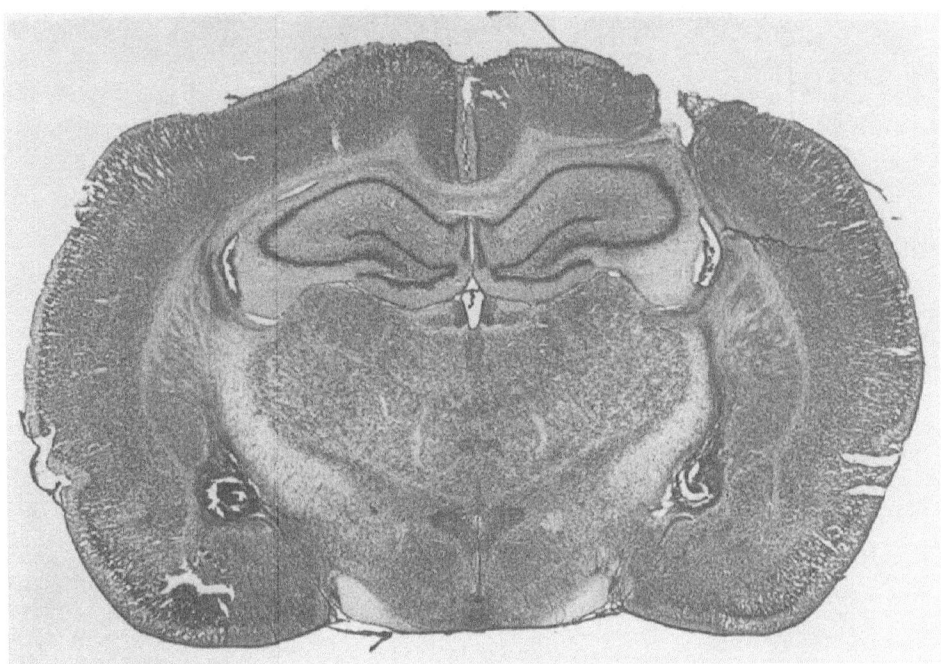

Fig. 5.14. Photomicrograph of central nucleus lesion (from Galaverna *et al.*, 1990).

HORMONAL INTERACTION OF ANGIOTENSIN AND ALDO IN THE BRAIN: CENTRAL NUCLEUS OF THE AMYGDALA

The central nucleus of the amygdala is known for its role in taste-motivated behaviors (Norgren, 1984). Recall that this site is rich in angiotensin receptors, and in angiotensinergic cells and fibers. It also contains ALDO receptors. Also recall from Chapter 3 that it is the major terminus of the taste–visceral pathway from the lower brainstem, and is known to be involved in cardiovascular regulation. The synergy of the response requires interactions between the effects of these hormones, perhaps at some distinct area. The central nucleus of the amygdala is a likely candidate (Chapter 3).

Functional studies corroborate this hypothesis. Lesions of this region (Fig. 5.14) interfere with intracerebrally delivered renin-induced sodium hunger, and DOCA-induced sodium hunger (Galaverna *et al.*, 1990; Fig. 5.15). Sodium depletion–induced salt intake is also impaired. The lesion, however, leaves water intake to the angiotensin intact. This means, for example, unlike lesions of the AV3V region which disrupt both the water and salt intake to centrally

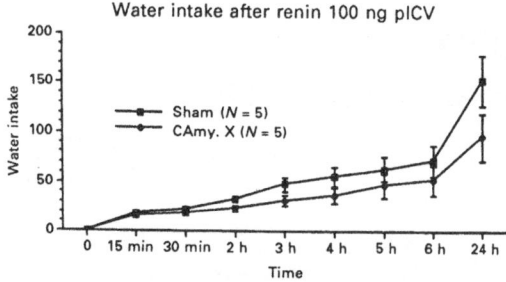

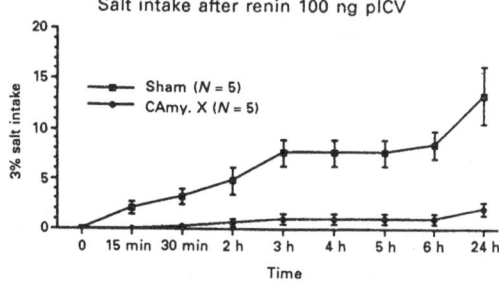

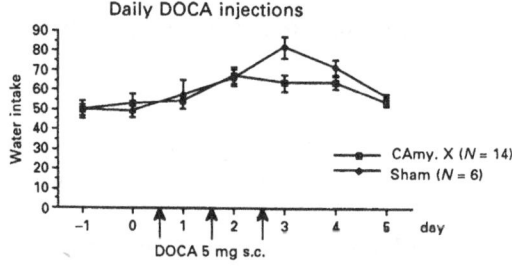

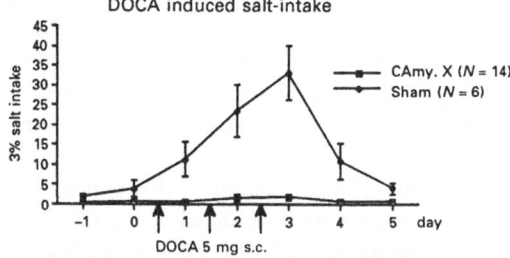

Fig. 5.15. Salt and water intake of lesion (Camy.X) and control (sham) rats to either ventricular intracerebral (pICV) injections of renin, or subcutaneous injections (s.c.) of DOCA (from Galaverna *et al.*, 1990).

delivered renin–angiotensin, central nucleus damage just interferes with the natriorexigenic effects. Thus, this region of the brain, so richly interconnected with all the major sites involved in body sodium and body fluid regulation, is essential for the natriorexigenic hormones to elicit their behavioral effects. Lesions of this site also impair other taste-motivated behavior, and even cardiovascular conditioning (Stellar & Stellar, 1985).

BRAIN REGIONS FOR ELEVATED SALT INTAKE

Chapter 1 described the fact that sodium-hungry rats increase their sodium intake on the second depletion more than on the first. One possible basis for enhancement of salt appetite in the multidepleted rat, and for its subsequent elevated need-free salt intake, is the hormonal organizational effects on the brain of angiotensin and ALDO in specific brain regions like the medial and central nucleus of the amygdala. For example, there may be irreversible changes in receptor function or in second messenger efficiency following the actions of these hormones on these brain regions. On the one hand, damage to the medial or central nucleus of the amgydala (Galaverna *et al.*, unpublished observations), or disconnection of the stria terminalis pathway (Black *et al.*, unpublished observations), or AV3V damage (DeLuca Jr *et al.*, unpublished observations) do not interfere with the enhancement of salt ingestion that appears on subsequent sodium depletion over the first one; the enhanced salt appetite, however, following sodium depletion treatments (see Chapter 1) is not elevated in neodecorticated rats (Schulkin & Grill, unpublished observations; Fig. 5.16) suggesting that neocortical tissue is involved in the phenomenon. On the other hand, damage to the central and medial nucleus of the amgydala reduces the salt ingestion that appears in the need-free NaCl intake following a history of sodium-depletion treatments (Nitabach *et al.*, 1989). By contrast, neither damage to the AV3V region (DeLuca Jr, *et al.*, unpublished observations), nor MPO damage (Michas *et al.*, unpublished observations), nor disconection of the stria terminalis pathway (Black *et al.*, unpublished observations) produce this effect.

INHIBITION OF SALT INTAKE

As has been discussed, sodium hunger is inhibited in the sheep by restoring sodium to the brain. In the rat, other factors are at work, some of which are specific to sodium hunger and other ingestive behaviors. Consider, first, four neurotransmitters, and brainstem sites in the satiation of the appetite.

It is clear that ANF is important in the regulation of extracellular fluid volume and sodium homeostasis: thirst and sodium ingestion (Chapters 2 and 4). Atrial peptide hormone has been localized in regions of the brain implicated in body fluid homeostasis and cardiovascular regulation (Chapters 2 and 4). By a combination of autoradiography (Gibson *et al.*, 1986; Bianchi *et al.*, 1986),

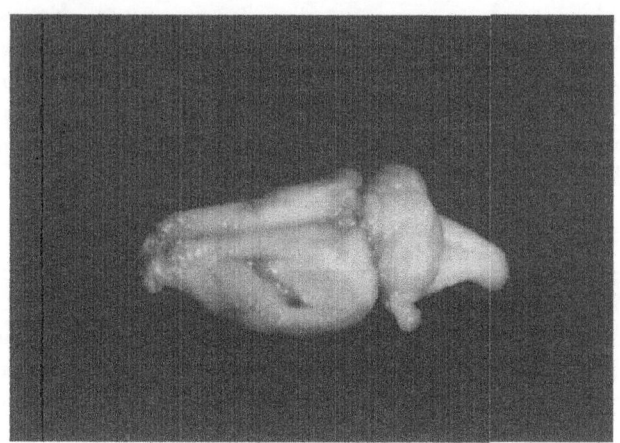

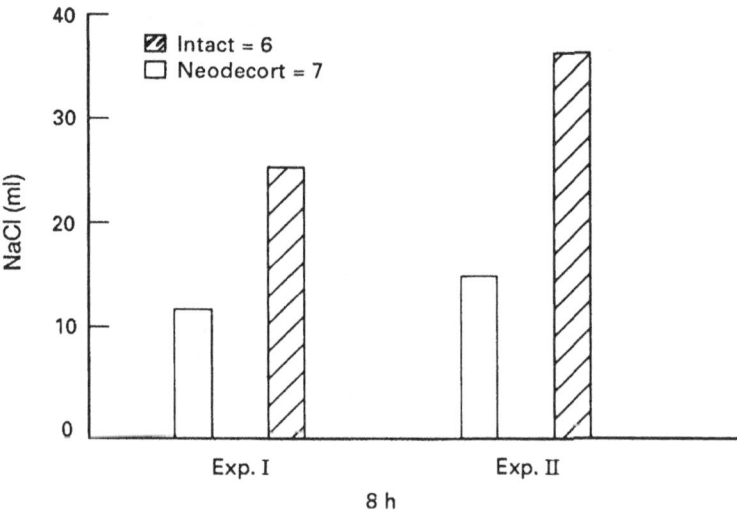

Fig. 5.16. Neodecorticated rat brain is depicted along with salt ingestion of sodium-deficient neodecorticated and control rats 8 h after being exposed to the salt (Schulkin & Grill, 1980).

immunoreactivity (Skofitsch *et al.*, 1985; Palkovits *et al.*, 1987) and ANF mRNA hybridization techniques, researchers have localized ANF in circumventricular organs, preoptic–hypothalamic tissue, and cardiovascular regions of the parabrachial and solitary nucleus (Jacobowitz *et al.*, 1985). As described in Chapter 2, intraventricular ANF reduces sodium depletion-induced salt appetite (e.g. Fitts *et al.*, 1985*a*) but does not reduce mineralocorticoid induced

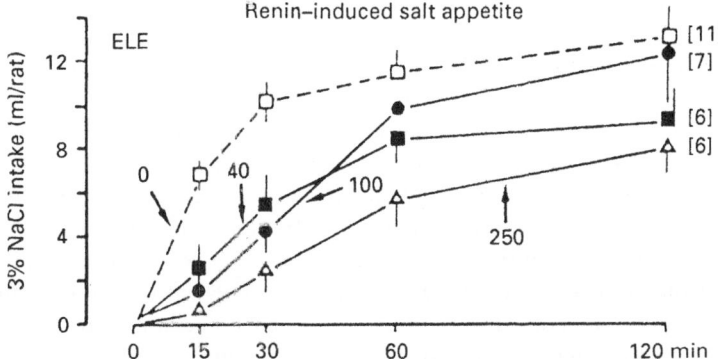

Fig. 5.17. Effects of bilateral injections into the medial amygdala of different doses (ng/rat) of eledoisin – a tachykinin (ELE) – on salt appetite induced by renin (from Massi *et al.*, 1990).

salt appetite (Massi & Schulkin, unpublished observations). This finding suggests that ANF inhibition of salt intake is on the angiotensinergic contribution to the genesis of salt intake. Other evidence indicates that ANF injected into the cerebral ventricles inhibits angiotensin-induced water and salt intake and reduces water deprivation-induced water drinking (as described in Chapter 2).

The tachykinins seem to be involved in body fluid homeostasis and blood pressure regulation; they have strong dipsogenic properties in the pigeon (DeCaro *et al.*, 1978; 1986), while inhibiting water intake when injected into the cerebral ventricles of rats (Massi *et al.*, 1988). They do not affect food intake (Massi *et al.*, 1988), but they do inhibit salt intake. That is, the tachykinins, eledosin, physalemin, kassinin and substance P inhibited sodium depletion-induced salt appetite (Massi *et al.*, 1988). Kassinin, the most potent inhibitor, reduced all forms of salt intake (Massi *et al.*, 1988). Moreover, eledosin, when injected into the cerebral ventricles, inhibited the salt intake of sodium–depleted sheep (McBurnie *et al.*, 1987), while actually increasing water intake. Sheep seem somewhere between the pigeon and the rat, in that water intake is increased but salt intake is decreased. The tachykinins localized in the medial amygdala, preoptic and bed nucleus of the stria terminalis also appear to affect the angiotensinergic contribution more so than the mineralocorticoid contribution. These same tachykinins, when injected into the amygdala, reduce depletion and renin-induced sodium hunger (e.g. Fig. 5.17), but do not affect DOCA-induced sodium hunger (Massi *et al.*, 1990). Therefore, the effects of the tachykinins, when injected into the medial amygdala, like ANF when injected into the cerebral ventricles, are selective for the angiotensinergic contribution in the genesis of a sodium hunger.

Oxytocin, which is localized in the paraventricular nucleus, has been shown to play a role in taste aversion learning and to inhibit salt intake in the rat (Stricker & Verbalis, 1987). It has been hypothesized, for example, that decreases in oxytocin synthesis (indicative of paraventricular nucleus activity) and secretion, result in the rat's ingestion of salt that results from adrenalectomy and DOCA. That is, the decrease of the oxytocinergic signal is excitatory for salt ingestion. Brain regions that synthesize the hormone, e.g. the paraventricular and supraoptic nuclei, are thought to contribute to the satiation of the appetite by decreasing the output of the neurohormone. The evidence for this hypothesis is that decreases in oxytocin levels are correlated with the onset and maintenance of the appetite; but oxytocin treatment does not decrease salt ingestion, nor is the appetite stimulated by oxytocin antagonists (Stricker & Verbalis, 1987; Stricker *et al.*, 1987). Moreover, the decrease in salt intake is not specific; the treatment is thought to provoke *malaise* or sickness; but perhaps the oxytocin contribution may be more closely coupled with its role in osmotic regulation.

Finally, cholinergic mechanisms are known to activate drinking behavior (reviewed by Rolls & Rolls, 1982), but may also be inhibitory on salt intake. In the brain, there is direct evidence that cholinergic signals inhibit salt intake. Carbachol, when injected into the third ventrical of the brain, inhibits salt intake of sodium-depleted rats; that inhibition also occurs when it is injected directly into the hypothalamus (e.g. Chiaraviglio & Taleisnik, 1969; Fitts *et al.*, 1985b). The inhibition of salt intake is specific; thirst, by contrast, is aroused by such treatments and food intake is not affected.

Perhaps, the above effects on salt intake are due to the activation of brainstem sites. It is known, more generally, that damage to the cardiovascular regions of the parabrachial nucleus (Ohman & Johnson, 1986), or the area postrema (Edwards & Ritter, 1982), result in elevated water ingestion to angiotensin or extracellular signals. And the area postrema is one region of the brain that is important in water and sodium balance (Contreras & Stetson, 1981; Hyde & Miselis, 1984). This region projects to the solitary nucleus and parabrachial region (e.g. Shapiro & Miselis, 1985). Area postrema neurons are catecholamine synthesizing and they project to cardiovascular regions of the medulla (Reis, 1981).

Aspirating the region of the area postrema and the caudal medial region of the solitary nucleus interferes with water balance (Contreras & Stetson, 1981; Hyde & Miselis, 1984; Schulkin, Hyde & Wolf, unpublished observations). The aspiration (Fig. 5.18) also results in elevated salt ingestion, and in an enhanced appetite compared with that of intact rats when rendered sodium deficient as shown in Fig. 5.18 (Uysal *et al.*, 1985). Area postrema-aspirated rats also tend to run faster for salt in a runway (described in Chapter 1) when mineralocorticoid

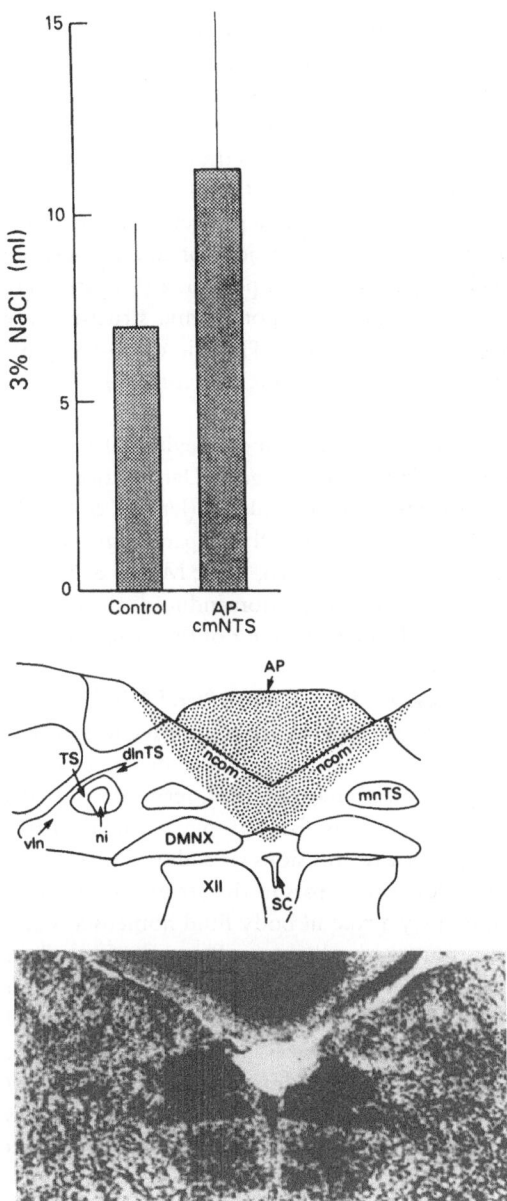

Fig. 5.18. Photomicrograph of aspirated area postrema (from Hyde & Miselis, 1984), along with the salt intake of sodium–depleted area postrema and control rats (from Uysal, Schulkin & Hyde, 1985).

treated (Arnell *et al.*, unpublished observations). Perhaps, then, this is one site involved in the satiation or inhibition of sodium hunger.

The area postrema and the caudal region of the solitary nucleus receive projections from the hepatic branch of the splanchnic nerve and the vagus nerve (e.g. Rogers *et al.*, 1984). Sites in the liver that project to the area postrema and other brainstem regions are known to play a role in sodium homeostasis. Both the heart and the liver project to the NTS and the area postrema via the vagus nerve (Shapiro & Miselis, 1985). Hepatic portal infusions of sodium activate caudal sites within the NTS which verge on the area postrema (Rogers *et al.*, 1979). Recall from Chapter 4 that changes in hypothalamic structures are known to occur following hepatic portal infusions. Hepatic osmo-receptors that influence the NTS also influence hypothalamic structures (Friedman, 1982).

The vagus nerve seems to play an important role in conveying information to the brain about sodium homeostasis from several organs. Denervation of the vagus nerve interferes with angiotensin-induced drinking (Jerome & Smith, 1982). And subdiaphragmatic vagotomized rats, as well as hepatic vagotomized rats, show decreased salt appetite (Contreras & Kosten, 1981; Martin & Novin, 1981; Tordoff *et al.*, 1986; Chapter 4) to sodium depletion-induced salt appetite, possibly because the lesion interferes with the processing of the vagal afferents that play a role in sodium homeostasis.

The hepatic infusion experiments described in Chapters 3 and 4 demonstrate that sodium-specific cells in the hepatic portal vein are responsive to levels of sodium in the hepatic portal system, and can result in the decrease of salt ingestion. Sites in the brain where hepatic afferents, in part, terminate overlap with gustatory regions of the brainstem; neurons, for example, within the PBN are known to change their firing pattern to infusions of sodium in the liver (Rogers *et al.*, 1979). In addition, volume receptors in the atrium of the heart monitoring blood flow and volume play a role in body fluid homeostasis and water and salt intake. The area postrema and solitary nucleus are informed, in part, by these actions, via the vagus nerve. Damage to the area postrema decreases cardiac output (Kosten *et al.*, 1983), while damage to the medial region of the solitary nucleus results in fulminating hypertension (Chapter 4). Hypothalamic regions that influence water and sodium balance (Stevenson *et al.*, 1950) and several cortical regions project to these caudal brainstem regions (Saper, 1982*a*, *b*), thereby directly influencing first-order cardiovascular regions of the brain.

Thus, while the evocation of salt ingestion requires forebrain sites, its termination via feedback from postabsorptive sites, perhaps, is mediated by hindbrain regions. One site is the caudal end of the solitary nucleus in addition to the area postrema. Hepatic atrial afferents are known to terminate in this region, and, therefore, it is ideally suited to monitor the consequences of the sodium ingestion.

BRAIN CIRCUITRY UNDERLYING SODIUM HUNGER

Recall that sodium hunger is a taste-guided behavior in which the gustatory system clearly plays an essential role (Chapter 3) and that the seventh, ninth, and tenth cranial nerves convey gustatory and more general visceral information to the solitary nucleus which, in turn, projects to the parabrachial region. Sectioning all three nerves abolishes sodium depletion-induced salt appetite. Lesions of the gustatory projections in the rostral part of the solitary nucleus disrupt it, as does damage to the parabrachial region. These results were described in Chapter 3, in addition to the fact that the parabrachial gustatory region sends a ventral projection, conveying taste and visceral information, through the reticular formation, zona incerta, and lateral hypothalamus to the amygdala and the bed nucleus of the stria terminalis; the parabrachial region also gives rise to a dorsal projection to the ventral medial thalamus and insular cortex. Various regions of the solitary nucleus also project to these sites in the ventral forebrain (Chapter 4) as do cardiovascular regions of the parabrachial region.

Lesions along the ventral taste–visceral pathway affect salt appetite. For example, mesencephalic reticular formation lesions at the level of the decussation of the brachium conjunctivum, ventrolateral to the central gray, or caudal to the thalamic gustatory region (Chiaraviglio, 1972), in addition to the lateral hypothalamic damage (Wolf, 1967), and damage to the zona incerta (Grossman & Grossman, 1978; Grossman, 1990), and whole amygdala damage (Cox *et al.*, 1978), all impair or abolish sodium depletion and mineralo-corticoid-induced salt appetite. It is interesting that neurons of the zona incerta are activated by the sight of sodium when sheep are hungry for sodium (Baldwin & Kendrick, 1988).

Lesions along the dorsal pathway do not disrupt salt appetite (Wolf *et al.*, 1970; Schulkin & Grill, 1980; Schulkin *et al.*, 1985*a, b*; e.g. thalamic taste region or insular cortex). Nor do lesions of the rostral, dorsomedial, medial, and caudal hypothalamus, the lateral preoptic region, olfactory bulbs, septum and dorsal hippocampus (Murphy & Brown, 1970; for review, see Wolf & Schulkin, 1980). Electrolytic lesions and knife cuts of the subfornical organ do not abolish sodium depletion or angiotensin-dependent sodium hunger (Schulkin *et al.*, 1983; Fitts & Mason, 1989; Thunhorst *et al.*, 1987; C.F. Weisinger *et al.*, 1990). These results suggest that it is the processing of information along the ventral taste–visceral pathway to the forebrain that is critical for the appetite (recall Chapter 3).

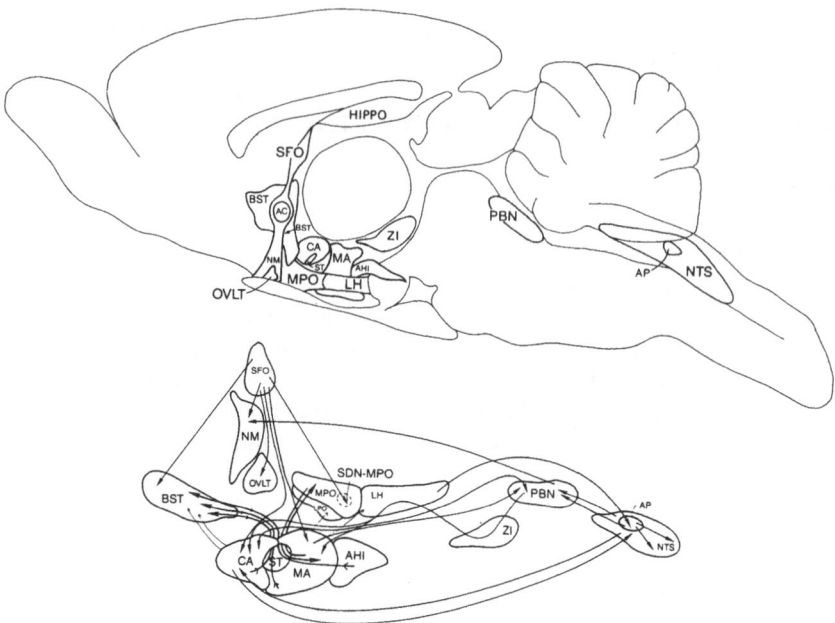

Fig. 5.19. Characterization of the rat brain showing (*above*) sites of special interest of a neural circuit which controls the behavioral expression of body sodium homeostasis, and (*below*) major connections among them AHI: amygdala–hippocampal regions; AC: anterior commisure; AP: area postrema; BST: bed nucleus of the stria terminalis; CA: central nucleus of the amygdala; Hippo: hippocampus; LH: lateral hypothalamus; LPO: lateral preoptic area; MA: medial nucleus of the amygdala; MPO: medial preoptic area; NM: nucleus medianus; NTS: nucleus of the tractus solitarius; OVLT; organum vasculosum of the lamina terminalis; PBN: parabrachial nucleus; SDN-MPO: sexually dimorphic nucleus of the medial preoptic area; SFO: subfornical organ; ST: stria terminalis; ZI: zona incerta (from Schulkin & Epstein, unpublished observations).

Figure 5.19 depicts a hypothetical neural circuit for salt appetite when aroused by angiotensin and ALDO. At the level of the brainstem, both taste and visceral information are processed in the solitary nucleus, area postrema and the parabrachial region. These brainstem sites project this information to the ventral forebrain where it is integrated with ALDO-sensitive regions (e.g. medial amygdala) or angiotensin-sensitive regions (median structures of the anterior ventral third ventricular region) and sites that are sensitive to both hormones (central nucleus of the amygdala). In other words, taste and visceral signals that carry information about sodium in the mouth or in the gastro-intestinal tract travel through ventral projections from the solitary nucleus and the parabrachial region to the forebrain (median structures of the anterior third ventricular tissue, preoptic area, bed nucleus of the stria terminalis and central and medial amygdala) where they reach separate brain regions that are sensitive to either angiotensin or ALDO, or both, and that are aroused by sodium depletion and which play a role in their synergistic action. Angiotensin appears to act within the anterior third ventricular region where it can respond to both blood-borne and central angiotensin. ALDO appears to act within the medial region of the amygdala, specifically within the medial nucleus. Both hormones seem to act on the central nucleus of the amygdala. These regions are interconnected, and the suggestion is that the synergistic effects of the two hormones requires communication between them.

Conclusion

The study of specific hungers, of which sodium hunger is one, dates at least in modern experimental form from the studies of Evvard (1915). He inquired into the food choices of swine during development, and found that they could select an adequate diet, when offered nine items, and grow well. Several years later, Osborne & Mendel (1918) demonstrated that rats, when offered choices of diets poor in protein and those rich in it, selected the rich diet. By 1925, Green showed that cattle were ingesting bone in response to phosphorus deficiency; he suggested that the ingestion of the bone was learned to ameliorate the phosphorus deficiency (see also Denton, 1982). The studies of Harris *et al.* (1933) encouraged the view that ingestion of diets in response to vitamin deficiencies were learned. But, there was a minority voice that began to suggest that animals used innate behavioral mechanisms in the selection of adaquate diets (e.g. Katz, 1937). Richter, as I have indicated, championed the idea that behavioral mechanisms mediating food and mineral choice were innate.

Sodium hunger, as it turns out, is paradigmatic of an innate specific hunger. This is one of the reasons why it is so interesting to study. In Chapter 1, it was demonstrated that the behavior of salt ingestion is innate, that there are hedonic changes in the perception of salt when rats are hungry for it, but also that significant learning interacts with innate mechanisms to guide the behavior.

But, there are other reasons why sodium hunger is interesting to study. One of the reasons is that the behavior of salt-seeking and its ingestion is generated in response to hormonal signals. Chapter 2 demonstrated that the behavior is, to a considerable degree, under hormonal control, and that the ingestion of salt is a sexually dimorphic behavior; females have a greater avidity for salt than do males. A reason for this sex difference may be found in the demands for minerals that females face during reproduction.

The search for salt when sodium hungry is guided by a specific sensory system: gustation. The gustatory system is the outer end of the alimentary tract for integrating what is out in the world with the demands of the internal milieu. The study of salt ingestion has played a major role in understanding how the gustatory system interacts with nutritional needs. Chapter 3 described the role of the gustatory system and its essential role in salt ingestion. I suggested that one reason why other mineral deficiencies result in the ingestion of NaCl is because the sensory quality of saltiness is correlated with the presence of other minerals. The one innate hunger for sodium is used to find sources of salt, and to contribute to the restoration of mineral imbalances perhaps other than sodium: because, at salt licks, a number of minerals are found together. It is the taste quality of saltiness that attracts animals to the salt licks, and is the marker for minerals.

As Chapter 4 has shown, the ingestion of sodium, along with water, is fundamental in the maintenance of extracellular fluid balance; the regulation of sodium is medically, in addition to physiologically, important. Many of the mechanisms of extracellular fluid regulation are also involved in cardiovascular control. The regulation of sodium involves many organs of the body including the heart, liver, adrenals, kidney, lung, pituitary, bone and brain. The study of sodium hunger highlights how many regions of the body work together in the regulation and control of this mineral.

But, it is the brain that is the primary organ for the expression of a sodium hunger. In Chapter 5, I have outlined a neural circuit that is responsive to the natriorexigenic hormones and that generates the motivated behavior. The neural circuit is also, to some extent, involved in other motivated behaviors, e.g. hunger, thirst and sex.

The study of sodium hunger is rich. Its analysis stretches from molecules to ecology. The fact that there are real species differences in the physiological and neurological substrates for eliciting the behavior cautions against any simple reduction, and is an invitation for comparative analysis. The analysis of sodium hunger is broad. The psychology has its roots in the biology of the animal. The behavior is easy to elicit and control in the laboratory. The result from the study of sodium hunger is how the brain produces a specific, motivated behavior.

Appendix

The research on sodium hunger is dependent upon methodological innovations in eliciting the behavior in the simplest, least distressful and most focused way for the animal. To this end, George Wolf devoted a fair amount of his research time trying to develop such simple procedures. The method he finally decided on was the combination of a sodium-deficient diet coupled with injections of furosemide (a diuretic and natriuretic agent). The treatment is usually over a one- or two-day period. The appetite appears within a day, probably even earlier. This is a simple acute procedure for eliciting the appetite and is now commonly used.

The route to this procedure traversed a number of techniques. One of the earliest was the use of formalin, a drug used in stress studies by Hans Selye. It provokes both hypovolemia and hyponatremia, but causes extensive damage to cells and capillaries; also, it is painful to the animal. Other methods still in use are polyethylene glycol or peritonal dialysis, which produces a hypovolemic animal. In this context, the animal is both thirsty and sodium hungry. Other methods include pharmacological doses of mineralocorticoids, or mineralocorticoids coupled with a sodium-deficient diet or with furosemide treatment. All of these treatments result in a robust salt appetite.

ACTH treatment, the combination of the hormones of sexual reproduction described in Chapter 2, elicits a salt appetite, in addition to angiotensin, or the synergy of angiotensin, and the mineralocorticoid hormones.

A technique used in the sheep has been the parotid fistula and the drainage of sodium from the gland. This results in a rather precise sodium loss that is commensurate with the sodium intake (Denton, 1982).

Finally, there are also a number of chronic treatments; adrenalectomy, chronic loss of sodium through the exteriorized parotid gland; being placed on a sodium-deficient diet on a long-term basis; or being implanted with a long-acting capsule that delivers mineralocorticoids are a few examples.

What the reader should appreciate is that the elicitation of a sodium hunger can be produced in a number of ways, is easy and reliable, and can be done in a virgin behavioral system.

References

Adam, W.R. & Dawborn, J.K. (1972). The effect of potassium depletion on mineral appetite in the rat. *Journal of Comparative and Physiological Psychology*, **78**, 51–8.

Aguilera, G. & Cait, K.J. (1983). Regulation of aldosterone secretion during altered sodium intake. *Journal of Steroid Biochemistry*, **19(1)**, 525–30.

Ahern, G., Landin, M.L. & Wolf, G. (1978). Escape frol deficits in sodium intake after thalamic lesions as a function of preoperative experience. *Journal of Comparative and Physiological Psychology*, **92**, 544–54.

Aldrich, E.C. (1939). Notes on the salt-feeding habits of the red crossbill. *The Condor*, **66**, 30.

Andersson, B., Dallman, M.F. & Olsson, K. (1969). Evidence for hypothalamic control of renal sodium excretion. *Acta Physiologia Scandinavia*, **75**, 496–510.

Andersson, B., Jobin, M. & Olsson, K. (1967) A study of thirst and other effects of an increased sodium concentration in the third brain ventricle. *Acta Physiologia Scandinavia*, **69**, 29–36.

Angulo, J.A, Schulkin, J. & McEwen, B.S. (1988). Effect of sodium depletion and aldosterone treatment on angiotensinogen mRNA in the brain of the rat. *Society for Neuroscience Abstracts*, **14**, 880.

Antunes-Rodriquez, J. & Covian, M.R. (1965). Specific changes in water intake and adipsia for water and sodium chloride after hypothalamic lesions. *Acta Physiologica Latin America*, **15**, 215–59.

Antunes-Rodriquez, J., Gentil, C.G., Negro-Vilar, A. & Covian, M.R. (1970). Role of adrenals in the changes of sodium chloride intake following lesions in the central nervous system. *Physiology and Behavior*, **5**, 89–94.

Antunes-Rodriquez, J., McCann, S.M. & Samson, W.K. (1986). Central administration of atrial natriuretic factor inhibits saline preference in the rat. *Endocrinology*, **118**, 1726–8.

Antunes-Rodriquez, J., Saad, W.A., Gentil, C.G. & Covian, M.R. (1970). Mechanism of decreased sodium chloride intake after hypothalamic lesions: effect of hydrochlorothiazide. *Physiology and Behavior*, **5**, 1183–6.

Arnold, A.P. & Breedlove, S.M. (1985). Organizational and activational effects of sex steroids on brain and behavior: a reanalysis. *Hormones and Behavior*, **19**, 469–98.

Arriza, J.L., Simerly, R.B., Swanson, L.W. & Evans, R.M. (1988). The neuronal mineralocorticoid receptor as a mediator of glucocorticoid response. *Neuron*, **42(1)**, 12–18.

Arriza, J.L., Weinberger, C., Cerelli, G., Glaser, T.M., Handelin, B.L., Housman, D.E. & Evans, R.M. (1987). Cloning of human mineralocorticoid receptor complementary DNA: structural and functional kinship with the glucocorticoid receptor. *Science*, **237**, 268–75.

Atarashi, K., Mulrow, P.J., Franco-Saenz, R., Snajdar, R. & Rapp, J. (1984). Inhibition of aldosterone production by an atrial extract. *Science*, **224**, 992–4.

Avrith, D.B. & Fitzsimons, J.T. (1980*a*). Increased sodium appetite in the rat induced by intracranial administration of components of the renin–angiotensin system. *Journal of Physiology*, **301**, 349–64.

Avrith, D.B., Wiselka, M.J. & Fitzsimons, J.T. (1980*b*). Increased sodium appetite in adrenalectomized or hypophysectomized rats after intracranial injections of renin or angiotensin II. *Journal of Endocrinology*, **87**, 109–12.

Baertschi, A.J. & Vallet, P.G. (1981). Osmosensitivity of the hepatic portal vein area and vasopressin release in rats. *Journal of Physiology*, **315**, 217–30.

Baldwin, B.A. (1968). The use of operant conditioning in the study of sodium appetite in goats. *Journal of Physiology*, **200**, 20–1.

Baldwin, B.A. & Kendrick, K.M. (1988). Sodium appetite modifies activity of cells in the zona incerta of sheep which respond to the sight or visual approach of food. *Proceedings of the Physiological Society*,

Balment, R.J., Brimble, M.J., Forsling, M.L. & Musabayane, C.T. (1985). Renal sodium retention in acutely hypophysectomized rats. *Journal of Physiology*, **361**, 46P.

Balment, R.J., Brimble, M.J., Forsling, M.L. & Musabayane, C.T. (1986). The influence of hormones on renal function in the acutely hypophysectomized rat. *Journal of Physiology*, **381**, 439–52.

Bare, J.K. (1949). The specific hunger for sodium chloride in normal and adrenalectomized white rats. *Journal of Comparative and Physiological Psychology*, **42(4)**, 242–53.

Barnwell, G.M., Dollahite, J. & Mitchell, D.S. (1985). Salt taste preference in baboons. *Physiology and Behavior*, **37**, 279–84.

Bartoshuk, L.M. (1964). Taste of sodium chloride solutions after adaptation to sodium chloride:Implications for the 'water taste'. *Science*, **143(3609)**, 967–8.

Bartoshuk, L.M. (1974). NaCl thresholds in man: thresholds for water taste or NaCl taste? *Journal of Comparative and Physiological Psychology*, **87(2)**, 310–25.

Bartoshuk, L.M. (1979). Preference changes: sensory vs hedonic explanations. In: J.H.A. Kroeze (ed.), *Preference Behavior and Chemoreception*. London: IRL.

Bartoshuk, L.M. (1980). Sensory analysis of the taste of NaCl. In: M. Kare, M. Fregly, and R. Bernard (eds), *Biological and Behavioral Aspects of Salt Intake*. New York: Academic Press.

Bartoshuk, L.M. (1988). Clinical psychophysics of taste. *Gerodontics*, **4**, 249–55.

Bartoshuk, L.M., Murphy, C. & Cleveland, C.T. (1978). Sweet taste of dilute NaCl: Psychophysical evidence for a sweet stimulus. *Physiology and Behavior*, **21**, 609–13.

Beach, F.A. (1971). Hormonal factors controlling the differentiation, and development, and display of copulatory behavior in the rat and related species. In: E. Toback, L. Aronson, and E. Shaw (eds) *Biopsychology of Development*, pp. 249–96, Academic Press.

Bealer, S.L. (1984). Hypothalamic knife cuts attenuate maintenance of deoxycorticosterone acetate – salt-induced hypertension. *Brain Research*, **309**, 192–5.

Bealer, S.L. & Johnson, A.K. (1979). Sodium consumption following lesions surrounding the anteroventral third ventricle. *Brain Research Bulletin*, **4**, 287–90.

Beatty, W.W. (1979). Gonadal hormones and sex differences in nonreproductive behaviors in rodents: 2ganization and activational influences. *Hormones and Behavior*, **12**, 112–63.

Beauchamp, G.K. (1987). The human preference for excess salt. *American Scientist*, **75**, 27–33.

Beauchamp, G.K. & Bertino, M. (1985). Rats (*Rattus norvegicus*) do not prefer salted solid food. *Journal of Comparative Psychology*, **99(2)**, 240–7.

Beauchamp, G.K., Bertino, M. & Engelman, K. (1983). Modification of salt taste. *Annals of Internal Medicina*, **98(2)**, 763–9.

Beauchamp, G.K., Bertino, M., Burke, D. & Engelman, K. (1990). Experimental sodium depletion and salt taste in normal human volunteers. *The American Journal of Clinical Nutrition*, **51**, 881–9.

Beauchamp, G.K., Cowart, B.J. & Moran, M. (1986). Developmental changes in salt acceptability in human infants. *Developmental Psychobiology*, **19(1)**, 17–25.

Beaumont, K., Vaughn, D.A. & Fanestil, D.D. (1987). Effect of adrenocorticoid receptors on potassium and sodium flux in rat C6 glioma cells. *Journal of Steroid Biochemistry*, **28(6)**, 593–8.

Beidler, L.M. (1954). A theory of taste stimulation. *Journal of General Physiology*, **38**, 133–9.

Beilharz, S. & Kay, R.N.B. (1963). The effects of ruminal and plasma sodium concentrations on the sodium appetite of sheep. *Journal of Physiology*, **165**, 468–83.

Beilharz, S., Bott, E.A., Denton, D.A. & Sabine, J.R. (1965). The effect of intracarotid infusions of 4M NaCl on the sodium drinking of sheep with a parotid fistula. *Journal of Physiology*, **178**, 80–91.

Bell, F.R. & Sly, J. (1979). The metabolic effects of sodium depletion in calves on salt appetite assessed by operant methods. *Journal of Physiology*, **295**, 431–43.

Bell, F.R. & Williams, H.L. (1960). The effect of sodium depletion on the taste threshold of calves. *Journal of Physiology*, **151**, 42–3.

Bell, F.R., Drury, P.L. & Sly, J. (1981). The effect on salt appetite and the renin–aldosterone system of replacing the depleted ions to sodium-deficient cattle. *Journal of Physiology*, **313**, 263–74.

Belovsky, G.E. (1978). Diet optimization in a generalist herbivore: the moose. *Theoretical Population Biology*, **14**, 105–134.

Ben-Ari, E.T. & Garrison, J.C. (1988). Regulation of angiotensinogen mRNA accumulation in rat hepatocyes. *Journal of Physiology*, E70–9.

Bergstrom, W.H. & Wallace, W.M. (1954). Bone as a sodium and potassium reservoir. *Journal of Clinical Investigation*, **33**, 867–73.

Berl, T. (1988). Psychosis and water balance. *New England Journal of Medicine*, **318(7)**, 441–2.

Bern, H.A. (1967). Hormones and endocrine glands of fishes, *Science*, **158**, 455–62.

Bernard, C. (1865, 1957). *An Introduction to the Study of Experimental Medicine*. New York: Dover Press.

Bernstein, I.L. (1988). Development of salt aversion in the Fischer-344 rat. *Developmental Psychobiology*, **21(7)**, 633–70.

Bernstein, I.L. & Courtney, L. (1987). Salt preference in the preweaning rat. *Developmental Psychobiology*, **20(4)**, 443–53.

Bernstein, I.L. & Hennessy, C.J. (1987). Amiloride-sensitive sodium channels and expression of sodium appetite in rats. *American Journal of Physiology*, R371–4.

Berridge, K.C. & Grill, H.J. (1983). Alternating ingestive and aversive consummatory responses suggest a two-dimensional analysis of palatability in rats. *Behavioral Neuroscience*, **97(4)**, 563–73.

Berridge, K.C. & Grill, H.J. (1984). Isohedonic tastes support a two-dimensional hypothesis of palatability. *Appetite*, **5**, 221–31.

Berridge, K.C. & Schulkin, J. (1989). Palatability shift of a salt-associated incentive drive during sodium depletion. *Quarterly Journal of Experimental Psychology*, **41B**, 121–38.

Berridge, K.C., Flynn, F.W., Schulkin, J. & Grill, H.J. (1984). Sodium depletion enhances salt palatability in rats. *Behavioral Neuroscience*, **98(4)**, 652–60.

Berridge, K.C., Grill, H.J. & Norgren, R. (1981). Relation of consummatory responses and preabsorptive insulin release to palability and taste aversions. *Journal of Comparative and Physiological Psychology*, **95**, 363–82.

Bertino, M. & Beauchamp, G.K. (1988). The spontaneously hypertensive rat's preference for salted foods. *Physiology and Behavior*, **44**, 285–9.

Bertino, M. & Chan, M.M. (1986). Taste perception and diet in individuals with Chinese and European ethnic backgrounds. *Chemical Senses*, **11(2)**, 229–41.

Bertino, M. & Tordoff, M.G. (1988). Sodium depletion increases rats' preferences for salted food. *Behavioral Neuroscience*, **102(4)**, 565–73.

Bertino, M., Beauchamp, G.K., Riskey, D.R. & Engelman, K. (1981). Taste perception in three individuals on a low sodium diet. *Appetite*, **2**, 67–73.

Bertorello, A., Hokfelt, T., Goldstein, M. & Aperia, A. (1988). Proximal tubule Na^+-K^+-ATPase activity is inhibited during high-salt diet: evidence for DA-mediated effect. *American Journal of Physiology*, **3**, F795–804.

Bianchi, C., Gutkowska, J., Ballack, M., Thibault, G., Garcia, R., Genest, J. & Cantin, M. (1986). Radioautographic localization of ^{125}I-atrial natriuretic factor binding sites in the brain. *Neuroendocrinology*, **44**, 365–72.

Biglieri, E.G. & Forsham, P.H. (1961). Studies on the expanded extracellular fluid and the responses to various stimuli in primary aldosteronism. *American Journal of Medicine*, **4**, 564–76.

Bird, E. & Contreras, R.J. (1986). Maternal dietary sodium chloride levels affect the sex ratio in rat litters. *Physiology and Behavior*, **36**, 307–10.

Birmingham, M.K., Sar, M. & Stumpf, W.E. (1986). Localization of aldosterone and corticosterone in the central nervous system assessed by quantitative autoradiography. *Neurochemical Research*, **9**, 333–50.

Black, R.M., Weingarten, H.P., Epstein, A.N., Maki, R. & Schulkin, J. (1990). The effects of disconnecting the stria terminalis on medial amygdala damage, on behavioral sodium regulation. *Physiology and Behavior*, in press.

Blair-West, J.R., Coghlan, J.P., Denton, D.A., Goding, J.R. & Wright, R.D. (1963). The effect of aldosterone, cortisol, and corticosterone upon the sodium and potassium content of sheep's parotid saliva. *Journal of Clinical Investigation*, **42(4)**, 484–96.

Blair-West, J.R., Coghlin, J.P., Denton, D.A., Scoggins, B.A., Wintour, M. & Wright, R.D. (1967). The renin–angiotensin–aldosterone system in sodium depletion. *Medical Journal of Australia*, **2**, 290.

Blair-West, J.R., Coghlan, J.P., Denton, D.A., Nelson, J.F., Orchard, E., Scoggins, B.A., Wright, R.D., Myers, K. & Junqueira, C.L. (1968). Physiological, morphological and behavioral adaptation to a sodium deficient environment by wild native Australian and introduced species of animals. *Nature, London* **217**, 922–8.

Blair-West, J.R., Denton, D.A., Gellatly, D.R., McKinley, M.J., Nelson, J.F. & Weisinger, R.S. (1987). Changes in sodium appetite in cattle induced by changes in CSF sodium concentration and osmolality. *Physiology and Behavior*, **39**, 465–9.

Blake, W.D. & Jurf, A.N. (1968). Increased voluntary Na intake in K deprived rats. *Communications in Behavioral Biology*, **A**, 1–7.

Blake, W.D. & Lin, K.K. (1978). Hepatic portal vein infusion of glucose and sodium solutions on the control of saline drinking in the rat. *Journal of Physiology*, **274**, 129–39.

Blass, E.M. & Hall, W.G. (1976). Drinking termination: interactions among hydrational, orogastric, and behavioral controls in rats. *Psychological Review*, **83(5)**, 356–74.

Bliss, D.K. & Bates, P.L. (1972). Choice intake of saline and water in pinealectomized rats under diurnal or constant lighting conditions. *Physiology and Behavior*, **9**, 429–33.

Bolles, R.C., Sulzbacher, S.I. & Arant, H. (1964). Innateness of the adrenalectomized rat's acceptance of salt. *Psychonomic Science*, **1**, 21–2.

Borer, K.T. (1968). Disappearance of preferences and aversions for sapid solutions in rats ingesting untasted fluids. *Journal of Comparative and Physiological Psychology*, **65(2)**, 213–21.

Botkin, D.B., Jordan, P.A., Dominski, A.S., Lowendorf, H.S. & Hutchinson, G.E. (1973). Sodium dynamics in a northern ecosystem. *Proceedings National Academy of Science, USA*, **70**, 2745–8.

Boudreau, J.C., Hoang, N.K., Oravec, J. & Do, L.T. (1983). Rat neurophysiological taste responses to salt solutions. *Chemical Senses*, **8(2)**, 131–50.

Boudreau, J.C., Sivakumar, L., Do, L.T., White, T.D., Oravec, J. & Hoang, N.K. (1985). Neurophysiology of geniculate ganglion (facial nerve) taste systems: species comparisons. *Chemical Senses*, **10(1)**, 89–127.

Brainard, J.B. (1976). Salt load as a trigger for migraine. *Minnesota Medicine*, **3**, 232–3.

Braun, J.J., Lasiter, P.S. & Kiefer, S.W. (1982). The gustatory neocortex of the rat. *Physiological Psychology*, **10(1)**, 13–45.

Braun-Menendez, E. (1952). Aumento del apetito especifico para la sal provocado por la desoxicorthicosterona. *Revista de la Sociedad Argentina de Biologia*, **XXVIII**, 15–23.

Braun-Menendez, E. (1953). Modificadores del apetito especifico para la sal en ratas blancas. *Revista de la Sociodad Argentina de Biologia*, **11**, 92–102.

Bregar, R.E., Strombakis, N., Allan, R.W. & Schulkin, J. (1983). Brief exposure to a saline stimulus promotes latent learning in the salt hunger system. *Neuroscience Abstracts*.

Brinton, R.E. & McEwen, B.S. (1988). Regional distinctions in the regulation of type 1 and type 11 adrenal steroid receptors in the central nervous system. *Neuroscience Research Communications*, **2(1)**, 37–45.

Brown, J.E. & Toma, R.B. (1986). Taste changes during pregnancy. *American Journal of Clinical Nutrition*, **43**, 414–18.

Bryant, R.W., Epstein, A.N., Fitzsimons, J.T. & Fluharty, S.J. (1980). Arousal of a specific and persistent sodium appetite in the rat with continuous intra-cerebroventricular infusion of angiotensin II. *Journal of Physiology*, **301**, 365–82.

Buggy, J. & Fisher, A.E. (1974). Water and sodium intake: evidence for a dual central role for angiotentensin. *Nature, London*, **250**, 733–5.

Buggy, J. & Johnson, A.K. (1977). Preoptic-hypothalamic periventricular lesions: thirst deficits and hypernatremia. *American Journal of Physiology*, R44–52.

Buggy, J. & Jonklaas, J. (1984). Sodium appetite decreased by central angiotensin blockade. *Physiology and Behavior*, **32**, 749–53.

Buggy, J., Huot, S., Pamnani, M. & Haddy, F. (1984). Periventricular forebrain mechanisms for blood pressure regulation. *Federation Proceedings*, **43**, 25–31.

Bunge, M. (1902). *Textbook of Physiology, Pathology & Chemistry*, pp. 82–103. London: Kegan Paul, Trench, Trubner and Company.

Burnell, G.M. & Foster, T.A. (1972). Psychosis with low sodium syndrome. Brief Communications, *American Journal of Psychiatry*, **128(10)**, 1313–14.

Bursey, R.G. & Watson, M.L. (1983). The effect of sodium restriction during gestation on offspring brain development in rats. *American Journal of Clinical Nutrition*, **1**, 43–51.

Butkus, A., Coghlan, J.P., Denton, D.A. & Scoggins, B.A. (1985). The effect of ACTH on the plasma concentrations of the 'hypertensinogenic' steroids, 17 x-hydroxyprogesterone and 17 α, 20 α-dihydroxy-4pregnen-3-one in sheep. *Journal of Steroid Biochemistry*, **22(3)**, 321–3.

Cabanac, M. (1971). Physiological role of pleasure. *Science*, **17(3)**, 1103–7.

Cabanac, M. (1979). Sensory pleasure. *Quarterly Review of Biology*, **54(1)**, 1–29.

Calaresu, F.R. & Ciriello, J. (1980). Projections to the hypothalamus from buffer nerves and nucleus tractus solitarius in the cat. *American Journal of Physiology*, R130–6

Camacho, A. & Phillips, M.I. (1981). Horseradish peroxidase study in rat of the neural connections of the organum vasculosum of the lamina terminalis. *Neuroscience Letters*, **25**, 201–4.

Cannon, W.B. (1915, 1929). *Bodily Changes in Pain, Hunger, Fear and Rage*. New York: Harper Torchbooks.

Cannon, W.B. (1932, 1966). *The Wisdom of the Body*. New York: Norton Press.

Cantin, M. & Genest, J. (1986). The heart as an endocrine gland. *Scientific American*, **4**, 76–81.

Carpenter, J.A. (1956). Species differences in taste preferences. *Journal of Comparative and Physiological Psychology*, **49(2)**, 139–44.

Carr, W.J. (1952). The effect of adrenalectomy upon the NaCl taste threshold in rat. *Journal of Comparative and Physiological Psychology*, **45**, 377–80.

Casto, R. & Phillips, M.I. (1985). Neuropeptide action in nucleus tractus solitarius: angiotensin specificity and hypertensive rats. *American Journal of Physiology*, **249**, R341–7.

Castren, E. & Saavedra, J.M. (1989). Angiotensin II receptors in paraventricular nucleus, subfornical organ, and pituitary gland of hypophysectomized, adrenalectomized, and vasopressin-deficient rats. *Neurobiology*, **86**, 725–9.

Catalanotto, F.A., Schechter, P.J. & Henkin, R.I. (1972). Preference for NaCl in the spontaneously hypertensive rat. *Life Sciences*, **11(1)**, 557–64.

Catalanotto, F.A. (1978). Effects of dietary methionine supplementation on preferences for NaCl solutions. *Behavioral Biology*, **24**, 457–66.

Chao, H.M., Choo, P.H. & McEwen, B.S. (1989). Glucocorticoid and mineralocorticoid receptor mRNA expression in rat brain. *Neuroendocrinology*, **50**, 365–71.

Chernigovsky, V.N. (1962). Morphophysiological structure of the interoceptive analyser and its role in the feeding behavior of animals. *Activitas Nervosa Superior*, **4**, 256–74.

Chiaraviglio, E. (1969). Effect of lesions in the septal area and olfactory bulbs on sodium chloride intake. *Physiology and Behavior*, **4**, 693–7.

Chiaraviglio, E. (1971). Amygdaloid modulation of sodium chloride and water intake in the rat. *Journal of Comparative and Physiological Psychology*, **76(3)**, 401–7.

Chiaraviglio, E. (1972). Mesencephalic influence on the intake of sodium chloride and water in the rat. *Brain Research*, 73–82.

Chiaraviglio, E. (1976a). Angiotensin–norepinephrine interaction on sodium intake. *Behavioral Biology*, **17**, 411–16.

Chiaraviglio, E. (1976b). Effects of renin–angiotensin system on sodium intake. *Journal of Physiology*, **255**, 57–66.

Chiaraviglio, E. (1984). Sodium chloride intake following electrochemical stimulation of the frontal lobe cortex in the rat. *Physiology and Behavior*, **33**, 547–51.

Chiaraviglio, E. & Perez Guaita, M.F. (1984). Anterior third ventricle (A3v) lesions and homeostasis regulation. *Journal of Physiology*, **79**, 446–52.

Chiaraviglio, E. & Perez Guaita, M.F. (1986). The effect of intracerebroventricular hypertonic infusion on sodium appetite in rats after peritoneal dialysis. *Physiology and Behavior*, **37**, 695–9.

Chiaraviglio, E. & Taleisnik, S. (1969). Water and salt intake induced by hypothalamic implants of cholinergic and adrenergic agents. *American Journal of Physiology*, **216(6)**, 1418–22.

Chimoskey, J.E., Spielman, W.S., Brandt, M.A. & Heidemann, S.R. (1984). Cardiac atria of BIO 14.6 hamsters are deficient in natriuretic factor. *Science*, **223**, 820–2.

Claire, M., Oblin, M-E, Steimer, J-L, Nakane, H., Misumi, J., Michaud, A. & Corvol, P. (1981). Effect of adrenalectomy and aldosterone on the modulation of mineralocorticoid receptors in rat kidney. *Journal of Biological Chemistry*, **256(1)**, 142–7.

Coirini, H., Magarinos, A.M., DeNicola, A.F., Rainbow, T.C. & McEwen, B.S. (1985). Further studies of brain aldosterone binding sites employing new mineralocorticoid and glucocorticoid receptor markers *in vitro*. *Brain Research*, **361**, 212–16.

Coirini, H., Marusic, E.T., DeNicola, A.F., Rainbow, T.C. & McEwen, B.S. (1983). Identification of mineralocorticoid binding sites in rat brain by competition studies and density gradient centrifugation. *Neuroendocrinology*, **37**, 354–60.

Coirini, H., Schulkin, J. & McEwen, B.S. (1988). Behavioral and neuroendocrine regulation of mineralocorticoid and glucocorticoid action. *Neuroscience Abstracts*.

Coleman, J.S., Fraser, J.D. & Pringle, C.A. (1985). *The Condor*, **87**, 291–2.

Contreras, R.J. (1977). Changes in gustatory nerve discharges with sodium deficiency: a single unit analysis. *Brain Research*, **121**, 373–8.

Contreras, R.J. (1979). Sodium deprivation alters neural responses to gustatory stimuli. *Journal of General Physiology*, **73**, 569–94.

Contreras, R.J. & Catalanotto, F.A. (1980). Sodium deprivation in rats: salt thresholds are related to salivary sodium concentrations. *Behavioral and Neural Biology*, **29**, 303–14.

Contreras, R.J. & Kosten, T. (1981). Changes in salt intake after abdominal vagotomy: Evidence for hepatic sodium receptors. *Physiology and Behavior*, **26**, 575–82.

Contreras, R.J. & Kosten, T. (1983). Prenatal and early postnatal sodium chloride intake modifies the solution preferences of adult rats. *Journal of Nutrition*, **113**, 1051–62.

Contreras, R.J. & Stetson, P.W. (1981). Changes in salt intake after lesions of the area postrema and the nucleus of the solitary tract in rats. *Brain Research*, **211**, 355–66.

Contreras, R.J., Bird, E. & Weisz, D.J. (1985). Behavioral and neural gustatory responses in rabbit. *Physiology and Behavior*, **34**, 761–8.

Contreras, R.J., Kosten, T. & Frank, M.E. (1984). Activity in salt taste fibers: peripheral mechanism for mediating changes in salt intake. *Chemical Senses*, **8(3)**, 275–88.

Cooke, H.J. (1988). Role of the 'little brain' in the gut in water and electrolyte homeostasis. *FASEB*, **3**, 127–38.

Correa, F. M.A., Plunkett, L.M. & Saavedra, J.M. (1986). Quantitative distribution of angiotensin-converting enzyme (Kininase 11) in discrete areas of the rat brain by autoradiography with computerized microdensitometry. *Brain Research*, **375**, 259–66.

Cort, J.H. (1963). Spontaneous salt intake in the rat following lesions in the posterior hypothalamus. *Physiologia Bohemoslovenica*, **12**, 502–5.

Covian, M.R. (1946). Apetito especifico de las ratas suprarrenoprivas para el cloruro de sodio. *Revista de la Sociedad Argentina de Biologia*, **22**, 383–93.

Covian, M.R. & Antunes-Rodrigues, J. (1963). Specific alterations in sodium chloride intake after hypothalamic lesions in the rat. *American Journal of Physiology*, **205(5)**, 922–6.

Covian, M.R., Gentil, C.G. & Antune-Rodrigues, J. (1972). Water and sodium chloride intake following microinjections of angiotensin 11 into the septal area of the rat brain. *Physiology and Behavior*, **9**, 373–7.

Cowan, I. & Brink, V.C. (1949). Natural game licks in the rocky mountain national parks of Canada. *Journal of Mammology*, **30**, 379–87.

Cowan, W.M., Raisman, G. & Powell, T.P.S. (1965). The connexions of the amygdala. *Journal of Neurology, Neurosurgery, and Psychiatry*, **28**, 137–50.

Cox, J.R., Cruz, C.E. & Ruger, J. (1978). Effect of total amygdalectomy upon regulation of salt intake in rats. *Brain Research Bulletin*, **3**, 431–5.

Crabbe, J. (1961). Stimulation of active sodium transport by the isolated toad bladder with aldosterone *in vitro*. *Journal of Clinical Investigation*, **40**, 2103–10.

Craig, W.C. (1918). Appetites and aversions as constituents of instincts. *Biological Bulletin*, **34**, 91–107.

Cruz, C., Perelle, I. & Wolf, G. (1977). Methodological aspects of sodium appetite: an addendum. *Behavioral Biology*, **20**, 96–103.

Cuello, A.C. & Kanazawa, I. (1978). The distribution of substance P immunoreactive fibers in the rat central nervous system. *Journal of Comparative Neurology*, **178**, 129–56.

Cullen, J.W. (1969). Modification of salt-seeking behavior in the adrenalectomized rat via gamma-ray irradiation. *Journal of Comparative and Physiological Psychology*, **68(4)**, 524–9.

Cullen, J.W. (1972). Sodium intake in the Mongolian gerbil (*Meriones unguiculatus*) consequent to subcutaneous formalin injections. *Psychonomic Science*, **26(5)**, 279–82.

Dahl, L.K. (1957). Evidence for an increased intake of sodium in hypertension based on urinary excretion of sodium. *Proceedings of the Society for Experimental Biology and Medicine*, **94**, 23–6.

Dahl, L.K. (1958). Salt intake and salt need. *New England Journal of Medicine*, **258**, 1152–7, 1205–8.

Dahl, L.K. & Love, R.A. (1954). Evidence for relationship between sodium (chloride) intake and human essential hypertension. *Archives of Internal Medicine*, **94**, 525–31.

Dahl, L.K., Heine, M. & Tassinari, L. (1962). Effects of chronic excess salt ingestion. Evidence that genetic factors play an important role in susceptibility to experimental hypertension. *Journal of Experimental Medicine*, **115(6)**, 1173–90.

Dahl, L.K. Heine, M. & Tassinari, L. (1963). High salt content of western infant's diet: possible relationship to hypertension in the adult. *Nature, London*, **198(4886)**, 1204–5.

Dalke, P.D., Beeman, R.D., Kindel, F.J., Robel, R.J., & Williams, T.R. (1965). Use of salt by elk in Idaho. *Journal of Wildlife Management*, **29**, 319–32.

Danielsen, J. & Buggy, J. (1980). Depression of *ad lib* and angiotensin-induced sodium intake at oestrus. *Brain Research Bulletin*, **5**, 501–4.

Darwin, (1872; 1965). *The Expression of the Emotions in Man and Animals*. University of Chicago Press, Chicago.

DeCaro, G., Epstein, A.N. & Massi, M. (1986). *The Physiology of Thirst and Sodium Appetite*. New York: Plenum Press.

DeCaro, G., Massi, M. & Micossi, L.G. (1978). Antidipsogenic effect of intracranial injections of substance P in rats. *Journal of Physiology*, **279**, 133–40.

Deems, R.O. & Friedman, M.L. (1988a). Sodium chloride preference is altered in a rat model of liver disease. *Physiology and Behavior*, **43**, 521–5.

Deems, R.O. & Friedman, M.L. (1988b). Altered preferences for sucrose, sodium chloride, urea and hydrochloric acid solutions in an animal model of cholestatic liver disease. *Physiology and Behavior*, **43**, 111–14.

DeKloet, E.R., Reul, J.M.H.M., DeRonde, F.S.W., Bloemers, M. & Ratka, A. (1986). Function and plasticity of brain corticosteroid receptor systems: action of neuropeptides. *Journal of Steroid Biochemistry*, **25(5B)**, 723–31.

DeLuca, Jr. L., Galaverna, O., Schulkin, J., & Epstein, A.N. (1990). Dependence of angiotensin-induced NaCl intake on the AV3V region. *Neuroscience Abstracts*.

DeLuca, L.A. Jr, Galaverna, O., Schulkin, J., Yao, S.Z. & Epstein, A.N. (1991). The anteroventral wall of the third ventricle and the angiotensinergic component of need-induced sodium intake in the rat. *Brain Research Bulletin*, in press.

DeNicola, A.F., Tornello, S., Coirini, H., Heller, C., Orti, E., White, A. & Marusic, E.T. (1982). Regulation of receptors for gluco- and mineralocorticoids. In:

Physiopathology of Hypophysial Disturbances and Diseases of Reproduction, New York: Alan R. Liss.

DeNicola, A.F., Tornello, S., Weisenberg, L., Fridman, O. Nd & Birmingham, M.K. (1981). Uptake and binding of (^{3}H) aldosterone by the anterior pituitary and brain regions in adrenalectomized rats. *Hormone Metabolism Research*, **13**, 103–6.

Denton, D.A. (1961). The selective appetite for Na$^+$ shown by Na$^+$ deficient sheep. *Journal of Physiology*, **157**, 97–116.

Denton, D.A. (1965). Evolutionary aspects of the emergence of aldosterone secretion and salt appetite. *Physiological Reviews*, **45(2)**, 245–95.

Denton, D.A. (1966). Some theoretical considerations in relation to innate appetite for salt. *Conditioned Reflex*, **1**, 144–70.

Denton, D.A. (1973). Sodium and hypertension. Mechanisms of hypertension. *Proceedings of an International Workshop Conference, Los Angeles, 7–10 March.* M.P. Sambhi, (ed.) Excerpta Medica, Amsterdam.

Denton, D.A. (1982). *The Hunger for Salt*. Berlin, Heidelberg, New York: Springer-Verlag.

Denton, D.A. (1984). Chemical and hormonal factors in the control of innate behaviour subserving ingestion. In *Frontiers in Physiological Research*, 335–47.

Denton, D.A. & Nelson, J.F. (1971). Effects of pregnancy and lactation on the mineral appetites of wild rabbits (*Oryctolagus cuniculus* (L.)). *Endocrinology*, **88**, 31–40.

Denton, D.A. & Sabine, J.R. (1963). The behaviour of Na deficient sheep. *Behaviour, XX*, **(3–4)**, 364–76.

Denton, D.A., Blair-West, J.R., McBurnie, M., Osborne, P.G. Tarjan, E., Williams, R.M., & Weisinger, R.S. (1990). Angiotensin and salt appetite of BALB/c mice. *American Journal of Physiology*, **28**, R729–35.

Denton, D.A., Coghlan, J.P., Fei, D.T., McKinley, M., Nelson, J., Scoggins, B., Tarjan, E., Tregear, G.W., Tresham, J.J., Weisinger, R. & Florey, H. (1984a). Stress, ACTH, salt intake and high blood pressure. *Clinical and Experimental Hypertension.* – Theory and Practice, **A6(1&2)**, 403–15.

Denton, D.A., McBurnie, M., Ong, F., Osborne, P. & Tarjan, E. (1988). Na deficiency and other physiological influences on voluntary Na intake of BALB/c mice. *American Journal of Physiology*, R1025–34.

Denton, D.A., McKinley, M.J., Nelson, J.F., Osborne, P., Simpson, J., Tarjan, E. & Weisinger, R.S. (1984b). Species differences in the effect of decreased CSF sodium concentration on salt appetite. *Journal of Physiology*, **69**, 499–504.

Denton, D.A., McKinley, M., Tarjan, E. & Weisinger, R. (1986, July). Sodium appetite: Some elements in an overview [Abstract]. *Proceedings of the Fourth International Conference on the Physiology of Food and Fluid Intake*, Seattle, Washington.

Denton, D.A., Nelson, J.F., Orchard, E. & Weller, S. (1969). The role of adrenocortical hormone secretion in salt appetite. *Olfaction and Taste*, 540–7.

Denton, D.A., Nelson, J.F. & Tarjan, E. (1985a). The voluntary correction of sodium deficiency by the rabbit. *Physiology and Behavior*, **34**, 181–7.

Denton, D.A., Nelson, J.F. & Tarjan, E. (1985b). Water and salt intake of wild rabbits (*Oryctolagus cuniculus*) (1) following dipsogenic stimuli. *Journal of Physiology*, **362**, 285–301.

Denton, D.A., Orchard, E. & Weller, S. (1970). The relation between voluntary sodium intake and body sodium balance in normal and adrenalectomized sheep. *Communications in Behavioral Biology*, **A(3)**, 213–21.

Deolmos, J.D., Alheid, G.F. and Beltramino, C.A. (1985). Amygdala. In: *The Rat Nervous System*, G. Pavinos (ed.) New York: Academic Press.

Deschepper, C.F., Bouhnik, J. & Ganong, W.F. (1986). Colocalization of angiotensinogen and glial fibrillary acidic protein in astrocytes in rat brain. *Brain Research*, **374**, 195–8.

Desor, J.A. & Maller, O. (1975). Preferences for sweet and salty in 9- to 15-year old and adult humans. *Science*, 686–7.

Desor, J.A., Maller, O. & Andrews, K. (1975). Ingestive responses of human newborns to salty, sour, and bitter stimuli. *Journal of Comparative and Physiological Psychology*, **89(8)**, 966–70.

Dethier, V.G. (1968). Chemosensory input and taste discrimination in the blowfly. *Science*, **161**, 389–91.

Deutsch, J.A. & Jones, A.D. (1960). Diluted water: An explanation of the rat's preference for saline. *Journal of Comparative and Physiological Psychology*, **53(2)**, 122–7.

Deutsch, J.A., Moore, B.O. & Heinrichs, S.C. (1989). Unlearned specific appetite protein. *Physiology and Behavior*, **46**, 619–24.

DeWardener, H.E. & Herxheimer, A. (1957). The effect of a high water intake on salt consumption, taste thresholds and salivary secretion in man. *Journal of Physiology*, **139**, 53–63.

Dickinson, A. (1980). *Contemporary Animal Learning Theory*. New York: Cambridge University Press.

Dickinson, A. (1986). Re-examination of the role of the instrumental contingency in the sodium-appetite irrelevant incentive effect. *Quarterly Journal of Experimental Psychology*, **38B**, 161–72.

Dickinson, A. & Nicholas, D.J. (1983*a*). Irrelevant incentive learning during training on ratio and interval schedules. *Quarterly Journal of Experiment Psychology*, **35B**, 235–47.

Dickinson, A. and Nicholas, D.J. (1983*b*). Irrelevant incentive learning during instrumental conditioning: the role of the drive–reinforcer and response–reinforcer relationships. *Quarterly Journal of Experimental Psychology*, **35B**, 249–63.

DiLorenzo, P.M. & Schwartzbaum, J.S. (1982). Coding of gustatory information in the pontine parabrachial nuclei of the rabbit: magnitude of neural response. *Brain Research*, **251**, 229–44.

DiNicolantonio, R., Hutchinson, J.S. & Mendelsohn, F.A.O. (1982). Exaggerated salt appetite of spontaneously hypertensive rats is decreased by central angiotensin-converting enzyme blockade. *Nature, London*, **298**, 846–8.

Dixon, J.S. (1958). Some biochemical aspects of deer licks. *Journal of Mammalology*, 109.

Donovick, P.J., Bliss, D.K., Burright, R.G. & Wertheim, L.M. (1973). Effect of pinealectomy or septal lesions on intake of unpalatable fluids in rats given sodium deplete or replete diets. *Physiology and Behavior*,(**10**, 1095–99.

Donovick, P.J. Burright, R.G. & Lustbader, S. (1970). Isotonic and hypertonic saline ingestion following septal lesions. *Communications in Behavioral Biology*, **4**, 17–22.

Douglas, J., Hansen, J. & Catt, K.J. (1978). Relationships between plasma rënin activity and plasma aldosterone in the rat after dietary electrolyte changes. *Endocrinology*, **103(1)**, 60–5.

Duncan, C.J. (1962). Salt preferences of birds and mammals. *Physiologic Zoology*, **35**, 120–32.

Duval, D. & Funder, J.W. (1974). The binding of tritiated aldosterone in the rat liver cytosol. *Endocrinology*, **94**, 575–9.

Edelman, I.S. (1978). Candidate mediates in the action of aldosterone on Na^+ transport. In *Membrane Transport Process* (J.F. Hoffman, ed.) vol. 1, New York: Raven Press.

Edelman, I.S., Bogoroch, R. & Porter, G.A. (1963). On the mechanism of action of aldosterone on sodium transport: The role of protein synthesis. *Proceedings of the National Academy of Sciences, USA*, **50(6)**, 1169–77.

Edwards, B.S., Zimmerman, R.S., Schwab, T.R., Heublein, D.M. & Burnett, J.C. Jr (1987). Atrial stretch, not pressure, is the principal determinant controlling the acute release of atrial natriuretic factor. *Circulation Research*, **62(2)**, 191–5.

Edwards, B.S., Burt, D., McIntyre, M.A., DeKloet, E.R., Stewart, P.M., Brett, L., Sutanto, W.S. & Monder, C. (1988). Localisation of 11B-hydroxysteroid dehydrogenase-tissue specific protector of the mineralocorticoid receptor. *Lancet*, 986–9.

Edwards, G.L. & Ritter, R.C. (1982). Area postrema lesions increase drinking to angiotensin and extracellular dehydration. *Physiology and Behavior*, **29**, 943–7.

Elfont, R.M., Epstein, A.N. & Fitzsimons, J.T. (1984). Involvement of the renin-angiotensin system in captopril-induced sodium appetite in the rat. *Journal of Physiology*, **354**, 11–27.

Elfont, R.M. & Sokol, H.W. (1987). Salt and water intake in brattleboro rats with hypothalamic diabetes inspidius. *Physiology and Behavior*, **39**, 53–61.

Emery, D.E. & Sachs, B.D. (1976). Copulatory behavior in male rats with lesions in the bed nucleus of the stria terminalis. *Physiology and Behavior*, **17**, 803–6.

Emson, P.C., Jessell, T., Paxinos, G. & Cuello, A.C. (1978). Substance P in the amygdaloid complex, bed nucleus and stria terminalis of the rat brain. *Brain Research*, **149**, 97–105.

Epstein, A.N. (1971). The lateral hypothalamic syndrome: Its implications for the physiological psychology of hunger and thirst. In: E. Stellar and J.M. Sprague (eds.). *Progress in physiological psychology*, vol. 4, New York: Academic Press.

Epstein, A.N. (1984). The dependence of the salt appetite of the rat on the hormonal consequences of sodium deficiency. *Journal of Physiology, Paris*, **79**, 494–8.

Epstein, A.N. & Massi, M. (1987). Salt appetite in the pigeon in response to pharmacological treatments. *Journal of Physiology*, **393**, 555–68.

Epstein, A.N. & Stellar, E. (1955). The control of salt preference in the adrenalectomized rat. *Journal of Comparative and Physiological Psychology*, **48(3)**, 167–72.

Epstein, A.N., Zhang, D-M., Schultz, J., Rosenberg, M., Kupsha, P. & Stellar, E. (1984). The failure of ventricular sodium to control sodium appetite in the rat. *Physiology and Behavior*, **32**, 683–6.

Erickson, R.P. anad Covey, E. (1980). On the singularity of taste sensations: what is a taste primary? *Physiology and Behavior*, **25**, 527–33.

Evans, R.M. (1988). The steroid and lyroid hormone receptor superfamily. *Science*, **240**, 889–95.

Evvard, J.M. (1915). Is the appetite of swine a reliable indication of physiological need? *Proceedings of the Iowa Academy of Science*, **22**, 375–402.

Falk, J.L. (1965a). Limitations to the specificity of NaCl appetite in sodium-depleted rats. *Journal of Comparative and Physiological Psychology*, 60(3(, 393–6.

Falk, J.L. (1966). Serial sodium depletion and NaCl solution intake. *Physiology and Behavior*, 1, 75–7.

Falk, J.L. (1965c). Water intake and NaCl appetite in sodium depletion. *Psychological Reports*, 315–25.

Falk, J.L. & Herman, T.S. (1961). Specific appetite for NaCl without postingestional repletion. *Journal of Comparative and Physiological Psychology*, 54(4), 405–8.

Feldstein, J.B., Sumners, C. & Raizada, M.K. (1986). Sodium increases angiotensin II receptors in neuronal cultures from brains of normotensive and hypertensive rats. *Brain Research*, 370, 265–72.

Ferrario, C.M., Gildenberg, P.L. & McCubbin, J.W. (1972). Cardiovascular effects of angiotensin mediated by the central nervous system. *Circulation Research*, XXX, 257–62.

Findlay, A.L.R. & Epstein, A.N. (1980). Increased sodium intake is somehow induced in rats by intravenous angiotensin II. *Hormones and Behavior*, 14, 86–92.

Findlay, A.L.R., Fitzsimons, J.T. & Kucharczyk, J. (1979). Dependence of spontaneous and angiotensin-induced drinking in the rat upon the oestrous cycle and ovarian hormones. *Journal of Endocrinology*, 82, 215–25.

Fisher, A.E. & Buggy, J. (1975). Central mediation of water and sodium intake: A dual role for angiotensin?, Section 7 Salt Appetite. In: *Control Mechanisms of Drinking*. G. Peters, J.T. Fitzsimons, and L. Peters-Haefeli. (eds) New York: Springer-Verlag.

Fitts, D.A. & Masson, D.B. (1989). Forebrain sites of action for drinking and salt appetite to angiotensin or captopril. *Behavioral Neuroscience*, 103, 865–72.

Fitts, D.A., Corp, E.S. & Simpson, J.B. (1982). Salt appetite and intravascular volume depletion following colloid dialysis in hamsters. *Behavioral and Neural Biology* 34, 75–88.

Fitts, D.A. & Mason, D.B. (1990). Preoptic angiotensin and salt appetite. *Behavioral Neuroscience*, 104, 643–50.

Fitts, D.A., Tjepkes, D.S. & Rochelle, O. (1990). Salt appetite and lesions of the ventral part of the ventral median preoptic nucleus. *Behavioral Neuroscience*, 103, 865–72.

Fitts, D.A., Thunhorst, R.L. & Simpson, J.B. (1985a). Diuresis and reduction of salt appetite by atrial ventricular infusions of atripeptin II. *Brain Research*, 332, 237–45.

Fitts, D., Thunhorst, R. & Simpson, J. (1985b). Modulation of salt appetite by lateral ventricular infusions of angiotensin II and carbachol during sodium depletion. *Brain Research*, 346, 273–80.

Fitts, D.A., Yang, O.O., Corp, E.S. & Simpson, J.B. (1983). Sodium retention and salt appetite following deoxycorticosterone in hamsters. *American Journal of Physiology*, R78–83.

Fitzsimons, J.T. (1961). Drinking by rats depleted of body fluid without increase in osmotic pressure. *Journal of Physiology*, 159, 297–309.

Fitzsimons, J.T. (1979). *The Physiology of Thirst and Sodium appetite*. Cambridge: Cambridge University Press.

Fitzsimons, J.T. & Fuller, L.M. (1985). Effects of angiotensin or carbachol on sodium intake and excretion in adrenalectomized or deoxycorticosterone-treated rats. *Journal of Physiology*, 359, 447–58.

Fitzsimons, J.T. & Moore-Gillon, M.J. (1980). Drinking and antidiuresis in response to reductions in venous return in the dog: neural and endocrine mechanisms. *Journal of Physiology*, **308**, 403–16.

Fitzsimons, J.T. & Stricker, E.M. (1971). Sodium appetite and the renin-angiotensin system. *New Biology*, **231(19)**, 58–60.

Fitzsimons, J.T. & Wirth, J.B. (1976). The neuroendocrinology of thirst and sodium appetite. In: *Central Nervous Control of Na$^+$ Balance-Relations to the Renin-Angiotensin System*. W. Kaufmann and D.K. Krause, (eds). Stuttgart: Georg Thieme Publishers.

Fluharty, S.J. & Epstein, A.N. (1983). Sodium appetite elicited by intracerebroventricular infusion of angiotensin II in the rat: II. Synergistic interaction with systemic mineralocorticoids. *Behavioral Neuroscience*, **97(5)**, 746–58.

Fluharty, S.J. & Manaker, S. (1983). Sodium appetite elicited by intracerebroventricular infusion of angiotensin II in the rat: 1. Relation to urinary sodium excretion. *Behavioral Neuroscience*, **97(5)**, 738–45.

Flynn, F.W. & Grill, H.J. (1988). Intraoral intake and taste reactivity responses elicited by sucrose and sodium chloride in chronic decerebrate rats. *Behavioral Neuroscience*, **102**, 934–41.

Flynn, F.W., Grill, H.J., Schulkin, J. & Norgren, R. (1992). Central gustatory lesions. II. Effects on sodium appetite, taste aversion learning and feeding behaviors, *Behavioral Neuroscience*, in press.

Fonberg, E. (1975). The amygdala and ingestive behavior. In: *Neural Mechanisms of Physiological Regulations and Behavior* G.R. Mogenson and F.R. Calaresu, (eds). Toronto: University of Toronto Press.

Forbes, G.B. & McCoord, A. (1965). Bone sodium as a function of serum sodium in rats. *American Journal of Physiology*, **209**, 830–4.

Formaker, B.K. & Hill, D.L. (1988). An analysis of residual NaCl taste response after amiloride. *American Journal of Physiology*, **R1002–R1007**.

Fox, N.A. (1985). Sweet/sour–interest/disgust: the role of approach – withdrawal in the development of emotions. In: *Infant Social Perception* T. Field and N.A. Fox (eds), Norwood, New Jersey: Ablex Press.

Fox, N.A. & Davidson, R.J. (1986). Taste-elicited changes in facial signs of emotion and the asymmetry of brain electrical activity in human newborns. *Neuropsychologia*, **24**, 417–22.

Frank, M.E. (1973). An analysis of hamster afferent taste nerve response functions. *Journal of General Physiology*, **61**, 588–618.

Frank, M.E. (1974). The classification of mammalian afferent taste nerve fibers. *Chemical Senses Flavor*, **1**, 53–60.

Frank, M.E. (1985). On the neural code for sweet and salty tastes. In: *Taste, Olfaction and the CNS* D. Pfaff (ed.), pp. 107–28. New York: Rockefeller University Press.

Frank, M.E., Contreras, R.J. & Hettinger, T.P. (1983). Nerve fibers sensitive to ionic taste stimuli in chorda tympani of the rat. *Journal of Neurophysiology*, **50(4)**, 941–60.

Frankmann, S.P., Sakai, R.R. & Simpson, J.B. (1987). Sodium appetite and cerebrospinal fluid sodium concentration during hypovolemia. *Appetite*, **9**, 57–64.

Fregly, M.J. (1958a). Specificity of the sodium chloride appetite of adrenalectomized rats; substitution of lithium chloride for sodium chloride. *American Journal of Physiology*, **195(3)**, 645–53.

Fregly, M.J. (1958b). NaCl appetite of adrenalectomized rats. *Proceedings of the Society for Experimental and Biological Medicine*, **97**, 144–8.

Fregly, M.J. (1973). Effect of an oral contraceptive on NaCl appetite and preference threshold in rats. *Pharmacology, Biochemistry and Behavior*, **1**, 61–8.

Fregly, M.J. (1980). Effect of chronic treatment with estrogen on the dipsogenic response of rats to angiotensin. *Pharmacology, Biochemistry and Behavior*, **12**, 131–6.

Fregly, M.J. & Taylor, R.E. Jr. (1964). Effect of hypothyroidism on water and sodium exchange in rats, In *Thirst in the Regulation of Body Water*, M.J. Wayner (ed.), Oxford: Pergamon Press, pp. 139–75.

Fregly, M.J., Harper Jr, J.H. & Radford, Jr, E.P. (1965). Regulation of sodium chloride intake by rats. *American Journal of Physiology*, **209**, 287–92.

Fregly, M.J. & Waters, I.W. (1966). Effect of mineralocorticoids on spontaneous sodium chloride appetite of adrenalectomized rats. *Physiology and Behavior*, **1**, 65–74.

Fregly, M.J. & Rowland, N.E. (1985). Role of renin–angiotensin–aldosterone system in NaCl appetite of rats. *American Journal of Physiology*, **248**, R1–R11.

Fregly, M.J. & Rowland, N.E. (1989). Preference threshold for NaCl solutions and sodium homeostasis in hypophysectomized rats. *Physiology and Behavior*, **45**, 825–9.

Friedman, M.I. (1982). Hepatic nerve function. In: *The Liver: Biology and Pathobiology*, I. Arias, H. Popper, D. Schacter, and D.A. Shafritz (eds) New York: Raven Press.

Friedman, M.I., Bruno, J.P. & Alberts, J.R. (1981). Physiological and behavioral consequences in rats of water recycling during lactation. *Journal of Comparative and Physiological Psychology*, **95(1)**, 26–35.

Fudim, O.K. (1978). Sensory preconditioning of flavors with a formalin-produced sodium need. *Journal of Experimental Psychology: Animal Behavior Processes*, **4(3)**, 276–85.

Funder, J.W. (1986). Adrenocorticoid receptors in the brain. In *Frontiers in Neuroendocrinology* (eds) W.F. Ganong and L. Martin vol. 9, Chap. 7, New York: Raven Press.

Funder, J.W. (1988). Mineralocorticoid action: Target tissue specificity is enzyme, not receptor, mediated. *Science*, **242**, 583–5.

Galaverna, O., DeLuca, L.A. Jr, Schulkin, J., Stellar, E. & Epstein, A.N. (1990). Deficits in NaCl ingestion after damage to the central nucleus of the amygdala in the rat, *Neuroscience Abstracts*.

Galef, B.G. Jr (1986). Social interaction modifies learned aversions, sodium appetite, and both palatability and handling-time induced dietary preference in rats (*Rattus norvegicus*). *Journal of Comparative Psychology*, **100(4)**, 432–9.

Ganchrow, J.R. & Erickson, R.P. (1970). Neural correlates of gustatory intensity and quality. *Journal of Neurophysiology*, **XXX111(6)**, 768–83.

Ganong, W.F. (1984). The brain renin–angiotensin system. *Annual Review Physiology*, **46**, 17–31.

Garcia, J., Hankins, W.G. & Rusiniak, K.W. (1974). Behavioral regulation of the milieu interne in man and rat. *Science*, **185**, 824–31.

Gardiner, T.W. & Stricker, E.M. (1985). Hyperdipsia in rats after electrolytic lesions of nucleus medianus. *American Journal of Physiology*, R214–23.

Gardiner, T.W., Jolley, F.R., Vagnucci, A.H. & Stricker, E.M. (1986). Enhanced sodium appetite in rats with lesions centered on nucleus medianus. *Behavioral Neuroscience*, **100(4)**, 531–5.

Genest, J. (1987). The brain, hypertension, and the atrial natriuretic factor. In: *Brain Peptides and Catecholamines in Cardiovascular Regulation*. J.P. Buckley and C.M. Ferrario, (eds) New York: Raven Press.

Gentil, C.G., Antunes-Rodrigues, J., Negro-Vilar, A. & Covian, M.R. (1968). Role of amygdaloid complex in sodium chloride and water intake in the rat. *Physiology and Behavior*, **3**, 981–5.

Gentil, C.G., Mogenson, G.J. & Stevenson, J.A.F. (1971). Electrical stimulation of septum, hypothalamus, and amygdala and saline preference. *American Journal of Physiology*, **220(5)**, 1172–7.

Gibbs, J., Younge, R.C. & Smith, G.P. (1973). Cholecystokin decreases food intake in rats. *Journal of Comparative and Physiological Psychology*, **84**, 488–95.

Gibson, J.J. (1966). *The Senses Considered as Perceptual Systems*. Boston: Houghton Mifflin Company.

Gibson, T.R., Wildey, G.M., Manaker, S. & Glembotski, C.C. (1986). Autoradiographic localization and characterization of atrial natriuretic peptide binding sites in the rat central nervous system and adrenal gland. *Journal of Neuroscience*, **6(7)**, 2004–11.

Gilbert, D.B. & Cooper, S.J. (1987). Effects of dopamine antagonists on fluid intake and salt preference in male and female rats. *Journal of Psychopharmacology*, **1**, 47–53.

Glaser, G.H. (1964). Sodium and seizures. *Epilepsia*, **5**, 97–111.

Gomez-Sanchez, E.P. (1986). Intracerebroventricular infusion of aldosterone induces hypertension in rats. *Endocrinology*, **118(2)**, 819–23.

Goodall, J. (1986). *The Chimpanzees of Gombe*. Calridge: Belknap Press of Harvard University Press.

Goy, R.W. & McEwen, B.S. (1977). *Sexual Differentiation of the Brain*. Cambridge: MIT Press.

Grace, J.E. (1968). Central nervous system lesions and saline intake in the rat. *Physiology and Behavior*, **3**, 387–93.

Grace, S.A. (1979). Plasma active and inactive renin in the rabbit: Effect of dietary sodium depletion and repletion. *Journal of Physiology*, **292**, 421–8.

Grace, S.A., Munday, K.A. & Noble, A.R. (1979). Sodium, potassium and water metabolism in the rabbit: the effect of sodium depletion and repletion. *Journal of Physiology*, **292**, 407–20.

Green, D.M., Coleman, D.H. & McCabe, M. (1948). Mechanisms of desoxycorticosterone action. II. Relation of sodium chloride intake to fluid exchange pressor effects and survival. **154**, 465–474.

Green, D.M., Saunders, F.J., Wahlgren, N., McDonough, F.J. & Clampit, J.M. (1952). Mechanisms of desoxycorticosterone action VII. Influence of the pituitary. *American Journal of Physiology*, **170(1)**, 107–15.

Green, H.H. (1925). Perceived appetites. *Physiology Reviews*, **5**, 336–48.

Grill, H.J. (1980). Production and regulation of ingestive consummatory behavior in the chronic decerebrate rat. *Brain Research Bulletin*, **5**, 79–87.

Grill, H.J. & Bernstein, I.L. (1988). Strain differences in taste reactivity to NaCl. *American Journal of Physiology*, R424–30.

Grill, H.J. & Miselis, R.R. (1979). Lack of ingestive compensation to dehydrational stimuli in decerebrates. *American Journal of Physiology*, **240**, R81–6.

Grill, H.J. & Norgren, R. (1978a). The taste reactivity test. I. Mimetic responses to gustatory stimuli in neurologically normal rats. *Brain Research*, **143**, 263–79.

Grill, H.J. & Norgren, R. (1978b). The taste reactivity test. II. Mimetic responses to gustatory stimuli in chronic thalamic and chronic decerebrate rats. *Behavioral Neuroscience*, **100**, 536–43.

Grimsley, D.L. (1968). Preoperative experience and postoperative fluid preference in adrenalectomized rats. *Psychonomic Science*, **12(3)**, 119–20.

Grimsley, D.L. (1970). Salt seeking in the adrenalectomized rat with an established sugar preference. *Journal of Comparative and Physiological Psychology*, **71(3)**, 384–9.

Grimsley, D.L. (1973). NaCl preference in the gerbil. *Physiological Psychology*, **1(1)**, 93–4.

Grossman, S.P. (1990). *Thirst and Sodium Appetite* New York: Academic Press.

Grossman, S.P. & Grossman, L. (1978). Parametric study of the regulatory capabilities of rats with rostromedial zona incerta lesions: Responsiveness to hypertonic saline and polyethylene glycol. *Physiology and Behavior*, **21**, 431–40.

Gutkind, J.S., Kurihara, M. & Saavedra, J.M. (1988). Increased angiotensin II receptors in brain nuclei of DOCA-salt hypertensive rats. *American Journal of Physiology*, H646–50.

Hall, C.E. & Hall, O. (1969). Interaction between desoxycorticosterone treatment, fluid intake, sodium consumption, blood pressure, and organ changes in rats drinking water, saline, or sucrose solution. *Canadian Journal of Physiology and Pharmacology*. **47(1)**, 81–6.

Halpern, B.P. (1983). Tasting and smelling as active, exploratory sensory processes. *American Journal of Otolaryngology*, **4**, 246–9.

Halpern, B.P. & Tapper, D.N. (1971). Taste stimuli: quality coding time. *Science*, 1256–8.

Hamlin, M.N., Webb, R.C., Ling, W.D. & Bohr, D.F. (1988). Parallel effects of DOCA on salt appetite, thirst, and blood pressure in sheep (42705). *Experimental Biology and Medicine*, **188**, 46–51.

Handal, P.J. (1965a). Formalin induced sodium appetite: dose–response relationships. *Psychonomic Science*, **3**, 511–12.

Handal, P.J. (1965b). Immediate acceptance of sodium salts by sodium deficient rats. *Psychonomic Science*, **3**, 315–16.

Harrelson, A. & McEwen, B. (1987). Steroid hormone influences on cyclic AMP-generating systems. *Current Topics in Membranes and Transport*, **31, 217–47.**

Harriman, A.E. (1955). The effect of a preoperative preference for sugar over salt upon compensatory salt selection by adrenalectomized rats. *Journal of Nutrition*, **57**, 271–6.

Harriman, A.E. (1967). Laughing gulls offered saline in preference and survival tests. *Physiological Zoology*, **40(3)**, 273–9.

Harriman, A.E. & MacLeod, R.B. (1953). Discriminative threshodds of salt for normal and adrenalectomized rats. *American Journal of Psychology*, **66**, 465–71.

Harris, L.J., Clay, J., Hargreaves, F. & Ward, A. (1933). Appetite and choice oings of the Royal Society of London, (Series B), **113**, 161–90.

Hartzell, A.K., Paulus, R.A. & Schulkin, J. (1985). Brief preoperative exposure to saline protects rats against behavioral impairments in salt appetite following central gustatory damage. *Behavioral Brain Research*, **15**, 9–13.

Hasegawa, H., Nasjletti, A., Rice, K. & Masson, G.M.C. (1973). Role of pituitary and adrenals in the regulation of plasma angiotensinogen. *American Journal of Physiology*, **225(1)**, 1–6.

Heck, G.L., Mierson, S. & DeSimone, J.A. (1984). Salt taste transduction occurs through an amiloride-sensitive sodium transport pathway. *Science*, **223**, 403–4.

Henkin, R.I. & Solomon, D.H. (1962). Salt taste threshold in adrenal insufficiency in man. *Journal of Clinical Endocrin Metabolism*, **22**, 856–8.

Henkin, R.I., Gill, J.R. Jr & Bartter, F.C. (1963). Studies on taste thresholds in normal man and in patients with adrenal cortical insufficiency: the role of adrenal cortical steroids and of serum sodium concentration. *Journal of Clinical Investigation*, **42(5)**, 727–35.

Herbert, D. & Cowan, I. McTaggart (1971). Natural salt licks as a part of the ecology of the mountain goat. *Canadian Journal of Zoology*, **49(5)**, 605–10.

Hermann, G.E., Kohlerman, N.J. & Rogers, R.C. (1983). Hepatic–vagal and gustatory afferent interactions in the brainstem of the rat. *Journal of the Autonomic Nervous System*, **9**, 477–95.

Herrick, C.J. (1905). The central gustatory paths in the brains of bony fishes. *Journal of Comparative Neurology and Psychology*, **15**, 375–456.

Herrick, C.J. (1948). *The Brain of the Tiger Salamander*, Ambystoma tigrinum. Chicago: University of Chicago Press.

Herxheimer, A. & Woodbury, D.M. (1960). The effect of deoxycorticosterone on salt sucrose taste preference thresholds and drinking behavior in rats. *Journal of Physiology*, **151**, 253–9.

Hilfehaus, M. (1976). Circadian rhythm of the renin–angiotensin – aldosterone system in the rat. *Archives of Toxicology*, **36**, 305–16.

Hill, D.L. (1987a). Development of taste responses in the rat parabrachial nucleus. *Journal of Neurophysiology*, **57(2)**, 481–95.

Hill, D.L. (1987b). Susceptibility of the developing rat gustatory system to the physiological effects of dietary sodium deprivation. *Journal of Physiology*, **393**, 413–24.

Hill, D.L. (1988). Influences of dietary sodium on functional taste receptor development: a sensitive period. *Science*, **241**, 1826–8.

Hill, D.L. & Almli, C.R. (1983). Parabrachial nuclei damage in infant rats produces residual deficits in gustatory preferences/aversions and sodium appetite. *Developmental Psychobiology*, **16(6)**, 519–33.

Hill, D.L. & Bour, T.C. (1985). Addition of functional amiloride-sensitive components to the receptor membrane: a possible mechanism for altered taste responses during development. *Developmental Brain Research*, **20**, 310–13.

Hill, D.L., Bradley, R.M. & Mistretta, C.M. (1983). Development of taste responses in rat nucleus of solitary tract. *Journal of Neurophysiology*, **50(4)**, 879–95.

Hill, D.L. & Mistretta, C.M. (1990). Developmental neuro-biology of salt taste sensation. *Trends in Neural Science*, **13**, 188–95.

Hill, D.L., Mistretta, C.M. & Bradley, R.M. (1986). Effects of dietary NaCl deprivation during early development on behavioral and neurophysiological taste responses. *Behavioral Neuroscience*, **100(3)**, 39–398.

Hill, D.L., Mistretta, C.M. & Bradley, R.M. (1982). Developmental changes in taste response characteristics of rat single chorda tympani fibers. *The Journal of Neuroscience*, **2(6)**, 782–90.

Hoffman, R.A. & Robinson, P.F. (1966). Changes in some endocrine glands of white-tailed deer as affected by season, sex and age. *Journal of Mammalogy*, **47(2)**, 266–80.

Horisberger, J-D. & Diezi, J. (1983). Effects of mineralocorticoids on Na^+ and K^+ excretion in the adrenalectomized rat. *American Journal of Physiology*, F89–99.

Horky, K. & Gregorova, I. (1980). Renin–angiotensin–aldosterone system in arterial hypertension. *Cor Vasa*, **22(1–2)**, 59–73.

Hoshishima, K., Yokoyama, S. & Seto, K. (1962). Taste sensitivity in various strains of mice. *American Journal of Physiology*, **202(6)**, 1200–4.

Hubel, D. and Weisel, T. (1979). Brain mechanisms of vision. *Scientific American*, **241**, 150–62.

Hull, C.L. (1943). *Principles of Psychology*. New York: Appelton-Century Croft.

Hwang, B.H., Chiueh, C.C. & Severs, W.B. (1984). Catecholamine synapses and contents in the paraventricular hypothalamic nucleus and nucleus tractus solitarius of DOCA-salt hypertensive rats. *The Anatomic Record*, **209**, 553–63.

Hyde, T.M. & Miselis, R.R. (1984). Area postrema and adjacent nucleus of the solitary tract in water and sodium balance. *American Journal of Physiology*, **247**, R173–82.

Ikonomov, O.C. Stoynev, A.G., Vrabchev, N.C., Shisheva, A.C. & Tarkolev, N.T. (1985). Circadian rhythms of food and 1% NaCl intake, urine and electrolyte excretion, plasma renin activity and insulin concentration in adrenalectomized rats. *Acta Physiologica Hungarica*, **65(2)**, 181–98.

Israel, A. & Barbella, Y. (1986). Diuretic and natriuretic action of rat atrial natriuretic peptide (6–33) administered intracerebroventricularly in rats. *Brain Research Bulletin*, **17**, 141–44.

Israel, A., Garrido, M.R., Barbella, Y. & Becemberg, I. (1988). Rat atrial natriuretic peptide (99–126) stimulates guanylate cyclase activity in rat subfornical organ and choroid plexus. *Brain Research Bulletin*, **20**, 253–6.

Iwao, H., Fukui, K., Kim, S., Nakayama, K., Ohkubo, H., Nakanishi, S. & Abe, Y. (1988). Sodium balance effects on renin, angiotensinogen, and atrial natriuretic polypeptide mRNA levels. *American Journal of Physiology*, E129–36.

Jackson, J.H. (1884, 1958). Evolution and dissolution of the nervous system. In: *Selected Writings of John Hughlings Jackson*. J. Taylor, (ed.) vol. 2 London: Staples Press.

Jacobowitz, D.M., Skofitsch, G., Keiser, H.R., Eskay, R.L. & Zamir, N. (1985). Evidence for the existence of atrial natriuretic factor-containing neurons in the rat brain. *Neuroendocrinology*, **40**, 92–4.

Jacobs, K.M., Mark, G.P. & Scott, T.R. (1988). Taste responses in the nucleus tractus solitarius of sodium-deprived rats. *Journal of Physiology*, **406**, 393–410.

Jakinovich, W. Jr & Osborn, D.W. (1981). Zinc nutrition and salt preference in rats. *American Journal of Physiology*, R223–9.

Jalowiec, J.E. (1974). Sodium appetite elicited by furosemide: effects of differential dietary maintenance. *Behavioral Biology*, **10**, 313–27.

Jalowiec, J.E. & Stricker, E.M. (1970a). Restoration of body fluid balance following acute sodium deficiency in rats. *Journal of Comparative and Physiological Psychology*, **70(1)**, 94–102.

Jalowiec, J.E. & Stricker, E.M. (1970*b*). Sodium appetite in rats after apparent recovery from acute sodium deficiency. *Journal of Comparative and Physiological Psychology*, **73(2)**, 238–44.

Jalowiec, J.E., Crapanzano, J.E. & Stricker, E.M. (1966). Specificity of salt appetite elicited by hypovolemia. *Psychonomic Science*, **6(7)**, 331–2.

Jalowiec, J.E., Stricker, E.M. & Wolf, G. (1970). Restoration of sodium balance in hypophysectomized rats after acute sodium deficiency. *Physiology and Behavior*, **5**, 1145–9.

Jerome, C. & Smith, G.P. (1982). Gastric or coeliac vagotomy decreases drinking after peripheral angiotensin II. *Physiology and Behavior*, **29**, 533–6.

Johnson, A.K. (1985). The periventricular anteroventral third ventricle (AV3V): its relationship with the subfornical organ and neural systems involved in maintaining body fluid homeostasis. *Brain Research Bulletin*, **15**, 595–601.

Johnston, J.B. (1923). Further contributions to the study of the evolution of the forebrain. *Journal of Comparative Neurology*, **35(5)**, 337–481.

Jones, R.L. & Hanson, H.C. (1985). *Mineral Licks, Geophagy, and Biogeochemistry of North American Ungulates*. Ames: Iowa State University Press.

Jonklaas, J. & Buggy, J. (1984). Angiotensin–estrogen interaction in female brain reduces drinking and pressor responses. *American Journal of Physiology*, R167–72.

Jonklaas, J. & Buggy, J. (1985). Angiotensin–estrogen central interaction: localization and mechanism. *Brain Research*, **326**, 239–49.

Kahrilas, P.J. & Rogers, R.C. (1984). Rat brainstem neurons responsive to changes in portal blood sodium concentration. *American Journal of Physiology*, R792–9.

Katz, D. (1937). *Animals and Men: Studies in Comparative Psychology*. London: Longmans.

Katzman, R. (1961). Central nervous system. Are glia high sodium cells? *Neurology*, **II**, 27–30.

Kaufman, S. (1980). A comparison of the dipsogenic responses of male and female rats to a variety of stimuli. *Canadian Journal of Physiology and Pharmacology*, **58(10)**, 1180–3.

Kaufman, S. (1981). Control of fluid intake in pregnant and lactating rats. *Journal of Physiology*, **318**, 9–16.

Kaufman, S. (1984). Role of right atrial receptors in the control of drinking in the rat. *Journal of Physiology*, **349**, 389–96.

Kaufman, S. & Mackay, B.J. (1983). Plasma prolactin levels and body fluid deficits in the rat: causal interactions and control of water intake. *Journal of Physiology*, **336**, 73–81.

Kaufman, S. & Monkton, E.A. (1988). Effect of peripherally administered atriopeptin III on water intake in rats. *Journal of Physiology*, **396**, 379–87.

Kaufman, S., Mackay, B.J. & Scott, J.Z. (1981). Daily water and electrolyte balance in chronically hyperprolactinaemic rats. *Journal of Physiology*, **321**, 11–19.

Kaunitz, H., Geller, L.M., Slanetz, C.A. & Johnson, R.E. (1960). Food restriction and salt preference in rats. *Nature, London*, **185**, 350–5.

Kelly, T.M. & Nelson, D.H. (1987). Sodium excretion and atrial natriuretic peptide levels during mineralocorticoid administration. A mechanism for the escape from hyperaldosteronism. *Endocrine Research*, **13(4)**, 363–83.

Kevetter, G.A. & Winans, S.S. (1981*a*). Connections of the corticomedial amygdala in the golden hamster. I. Efferents of the 'vomeronasal amygdala'. *Journal of Comparative Neurology*, **197**, 81–98.

Kevetter, G.A. & Winans, S.S. (1981*b*). Connections of the corticomedial amygdala in the golden hamster. II. Efferents of the 'olfactory amygdala'. *Journal of Comparative Neurology*, **197**, 99–111.

Khalil, K.A. & Eisman, E.H. (1971). Some parameters of latent learning and generalized drives. *Journal of Comparative and Physiological Psychology*, **77(3)**, 463–469.

Kiefer, S.W. (1978). Two-bottle discrimination of equimolar NaCl and LiCl solutions by rats. *Physiological Psychology*, **6(2)**, 191–8.

King, S.J., Harding, J.W. & Moe, K.E. (1988). Elevated salt appetite and brain binding of angiotensin II in mineralocorticoid-treated rats. *Brain Research*, **448**, 140–9.

Kissileff, H.R. & Epstein, A.N. (1962). Loss of salt preference in rats with lateral hypothalamic damage. *American Zoologist*, **2(4)**, 116.

Kissileff, H.R. & Hoeffer, R. (1975). Reduction of saline intake in adrenalectomized rats during chronic intragastric infusions of saline. In: *Control Mechanisms of Drinking*, G. Peter, J.T. Fitzsimons, and L. Peters-Haefeli (eds). Heidelberg: Springer-Verlag.

Knuepfer, M.M., Johnson, A.K. & Brody, M.J. (1984). Identification of brainstem projections mediating hemodynamic responses to stimulation of the anteroventral third ventricle (AV3V) region. *Brain Research*, **294**, 305–14.

Koh, S.D. & Teitelbaum, P. (1961). Absolute behavioral taste thresholds in the rat. *Journal of Comparative and Physiological Psychology*, **54(3)**, 223–9.

Kosar, E., Grill, H.J. & Norgren, R. (1986). Gustatory cortex in the rat. I. Physiological properties and cytoarchitecture. *Brain Research*, **379**, 329–41.

Kosten, T. & Contreras, R.J. (1985). Adrenalectomy reduces peripheral neural responses to gustatory stimuli in the rat. *Behavioral Neuroscience*, **99(4)**, 734–41.

Kosten, T., Contreras, R.J., Stetson, P.W. & Ernest, M.J. (1983). Enhanced saline intake and decreased heart rate after area postrema ablations in rat. *Physiology and Behavior*, **31**, 777–85.

Kraulis, I., Foldes, G., Traikov, H., Dubrovsky, B. & Birmingham, M.K. (1975). Distribution, metabolism and biological activity of deoxycorticosterone in the central nervous system. *Brain Research*, **88**, 1–14.

Krecek, J. (1973). Sex differences in salt taste: the effect of testosterone. *Physiology and Behavior*, **10**, 683–8.

Krecek, J. (1975). The pineal gland and the development of salt intake patterns in male rats. *Developmental Psychobiology*, **9(2)**, 181–8.

Krecek, J. (1978). Effect of ovarectomy of females and oestrogen administration to males during the neonatal critical period on salt intake in adulthood in rats. *Physiologia Bohemoslovaca*, **27**, 1–5.

Krecek, J., Novakova, V. & Stibral, K. (1972). Sex differences in the taste preference for a salt solutionon in the rat. *Physiology and Behavior*, **8**, 183–8.

Krecek, J., Panek, M., Salatova, J. & Zicha, J. (1975). The pineal gland and the effect of neonatal administration of androgen upon the development of spontaneous salt and water intake in female rats. *Neuroendocrinology*, **18**, 137–43.

Krettek, J.E. & Price, J.L. (1978). Amygdaloid projections to subcortical structures within the basal forebrain and brainstem in the rat and cat. *Journal of Comparative Neurology*, **178**, 225–54.

Krieckhaus, E.E. (1970). 'Innate recognition' aids rats in sodium regulation. *Journal of Comparative and Physiologica, Psychology*, **73(1)**, 117–22.

Krieckhaus, E.E. and Wolf, G. (1968). Acquisition of sodium by rats: interaction of innate mechanisms and latent learning. *Journal of Comparative and Physiological Psychology*. **65(2)**, 197–201.

Krozowski, Z.S. & Funder, J.W. (1981). Mineralocorticoid receptors in rat anterior pituitary: toward a redefinition of 'mineralocorticoid hormone'. *Endocrinology*, **109(4)**, 1221–4.

Kucharczyk, J. (1984*a*). Localization of central nervous system structures mediating extracellular thirst in the female rat. *Journal of Endocrinology*, **100**, 183–8.

Kucharczyk, J. (1984*b*). Neuroendocrine mechanisms mediating fluid intake during the estrous cycle. *Brain Research Bulletin*, **12**, 175–80.

Kurtz, T.W. & Morris, R.C. Jr (1985). Hypertension and sodium salts. *Science*, **228**, 352–3.

Kuta, C.C., Bryant, H.V., Zabik, J.E. & Yim, Gik. W. (1984). Stress, endogenous opioids and salt intake. *Appetite*, **5**, 53–60.

Kutscher, C.L. & Steilen, H. (1973). Increased drinking of a nonpreferred NaCl solution during food deprivation in the C3H/HeJ mouse. *Physiology and Behavior*, **10**, 29–34.

Lan, N.C., Matulich, D.T., Morris, J.A. & Baxter, J.D. (1981). Mineralocorticoid receptor-like aldosterone-binding protein in cell culture. *Endocrinology*, **109(6)**, 1963–70.

Lassman, M.N. & Mulrow, P.J. (1974). Deficiency of deoxycorticosterone-binding protein in the hypothalamus of rats resistance to deoxycorticosterone-induced hypertension. *Endocrinology*, **94**, 1541–6.

Lautt, W.W. (1983). Afferent and efferent neural roles in liver function. *Progress in Neurobiology*, **21**, 323–48.

Leaf, A. (1984). Dehydration in the elderly. *New England Journal of Medicine*, Editorial. **311(12)**, 791–2.

Lehman, M.N. & Winans, S.S. (1980). Medial nucleus of the amygdala mediates chemosensory control of male hamster sexual behavior. *Science*, **210**, 557–60.

LeMagnen, J. (1985). *Hunger*. New York: Cambridge University Press.

Lemoine, J. and Kucharczyk, J. (1985). Fluid regulation and reproductive cyclicity in female rats treated neonatally with testosterone and methandrostenolone. *Hormone Research*, **22**, 291–300.

Leshem, M. & Epstein, A.N. (1989). Ontogeny of renin-induced salt appetite in the rat pup. *Developments in Psychobiology*, **22**, 437–45.

Levy, C.J. & McCutcheon, B. (1974). Importance of postingestional factors in the satiation of sodium appetite in rats. *Physiology and Behavior*, **13**, 621–5.

Levy, M. & Wexler, M.J. (1987). Hepatic denervation alters first-phase urinary sodium excretion in dogs with cirrhosis. *American Journal of Physiology*, F664–71.

Lewis, M. (1960). Behavior resulting from sodium chloride deprivation in adrenalectomized rats. *Journal of Comparative and Physiological Psychology*, **53(5)**, 464–7.

Lewis, M. (1968). Discrimination between drives for sodium chloride and calcium. *Journal of Comparative and Physiological Psychology*, **65(2)**, 208–12.

Lin, K-K. & Blake, W.D. (1971). Hepatic sodium receptor in control of saline drinking behavior. *Communications in Behavioral Biology*, **5**, 359–63.

Lind, R.W. (1988). Sites of action of angiotensin in the brain. In *Angiotensin and Blood Pressure Regulation*, J. Harding, J. Wright, R.C. Speth, & N. Barnes (eds). New York: Academic Press.

Lind, R.W., Swanson, L.W. & Ganten, D. (1985). Organization of angiotensin II immunoreactive cells and fibers in the rat central nervous system. *Neuroendocrinology*, **40**, 2–24.

Lynch, K.R., Hawelu-Johnson, C.L., & Guyenet, P.G. (1987). Localization of brain angiotensinogen mRNA by hybridization histochemistry. *Molecular Brain Research*, **2**, 149–58.

McBurney, D.H. & Pfaffmann C. (1963). Gustatory adaptation to saliva and sodium chloride. *Journal of Experimental Psychology*, **6**, 523–9.

McBurnie, M., Denton, D. & Tarjan, E. (1988). Influence of pregnancy and lactation on Na appetite of BALB/c mice. *American Journal of Physiology*, R1020–4.

McBurnie, M., Decaro, G., Denton, D.A., Massi, M. & Tarjan, E. (1987). Effect of ICV infusion of eledoisin on the sodium and water intake of sheep. Abstract, Integrative Mechanisms in Neural Function, *Swedish Australian Science Symposium*, Melbourne, Australia.

McCance, R.A. (1936). Medical problems in mineral metabolism: III. Experimental human salt deficiency. *Lancet*, **230**, 823–30.

McCance, R.A. (1938). The effect of salt deficiency in man on the volume of the extracellular fluids and on the composition of sweat, saliva, gastric juice and cerebrospinal fluid. *Journal of Physiology*, **92**, 208–18.

McCleary, R.A. (1953). Taste and post-ingestion factors in specific-hunger behavior. *Journal of Comparative Physiological Psychology*, **46**, 411–20.

McCutcheon, B. & Levy, C. (1972). Relationship between NaCl rewarded bar-pressing and duration of sodium deficiency. *Physiology and Behavior*, **8**, 761–3.

McEwen, B.S. (1976). Interactions between hormones and nerve tissue. *Scientific American*, **235**, 48–58.

McEwen, B.S. (1989). Endocrine effects on the brain and their relationship to behavior. In: *Basic Neurochemistry: Molecular Cellular and Medical Aspects*, 4th edn G.J. Siegel. (ed.) New York: Raven Press.

McEwen, B.S. & Pfaff, D.W. (1985). Hormone effects on hypothalamic neurons: analysing gene expression and neuromodulator action. *Trends in Neurological Science Reviews*, March, 105–10.

McEwen, B.S., Jones, K.J. & Pfaff, D.W. (1987). Hormonal control of sexual behavior in the female rat: molecular, cellular and neurochemical studies. *Biology of Reproduction*, **36**, 37–45.

McEwen, B.S., Lambin, L.T., Rainbow, T.C. & DeNicola, A.F. (1986). Aldosterone effects on salt appetite in adrenalectomized rats. *Neuroendocrinology*, **43**, 38–43.

McKenzie, J.S. & Denton, D.A. (1974). Salt ingestion responses to diencephalic electrical stimulation in the unrestrained conscious sheep. *Brain Research*, **70**, 449–66.

McKinley, M.J., Allen, A., Denton, D.A., Clevers, J., Mendelsohn, F.A.O. & Tarjan, E. (1986*a*). Localization of angiotensin II receptors in rabbit and sheep brain by *in vitro* autoradiography. *Appetite*, **7**, 280.

McKinley, M.J., Blaine, E.H. & Denton, D.A. (1974). Brain osmoreceptors, cerebrospinal fluid electrolyte composition and thirst. *Brain Research*, **70**, 532–7.

McKinley, M.J., Denton, D.A., Coghlan, J.P., Harvey, R.B., McDougall, J.G., Rundgren, M., Scoggins, B.A. & Weisinger, R.S. (1986*b*). Cerebral osmoregulation of renal sodium excretion – a response analogous to thirst and vasopressin release. *Canadian Journal of Physiological Pharmacology*, **65**, 1724–9.

McMurray, T.M. & Snowdon, C.T. (1977). Sodium preferences and responses to sodium deficiency in rhesus monkeys. *Physiological Psychology*, **5(4)**, 477–82.

Maes, F.W. & Erickson, R.P. (1984). Gustatory intensity discrimination in rat NTS: a tool for the evaluation of neural coding theories. *Journal of Comparative Physiology A*, **155**, 271–82.

Magarinos, A.M., Coirini, H., DeNicola, A.F. & McEwen, B.S. (1986). Mineralocorticoid regulation of salt intake is preserved in hippocampectomized rats. *Neuroendocrinology*, **44**, 494–7.

Magliola, L., McMahon, E.G. & Jones, A.W. (1986). Alterations in active Na–K transport during mineralocorticoid-salt hypertension in the rat. *American Journal of Physiology*, C540–44.

Mann, H., Stiller, S. & Korz, R. (1976). Biological balance of sodium and potassium. *Pflugers Archiv, European Journal of Physiology*, **362**, 135–9.

Marini, J., Schulkin, J. & Epstein, A.N. (1986). The role of the medial region of the amygdala in aldosterone-induced salt appetite. *Neuroscience Abstract*.

Marler, P.R. & Hamilton, W.J. (1966). *Mechanisms of Animal Behavior*. New York: John Wiley.

Martin, J.R. & Novin, D. (1981). Response to dipsogenic stimuli after abdominal vagotomy in rats. *Physiological Psychology*, **9(2)**, 181–6.

Martin, R.L. & Hammond, G.E. (1983). Lateral hypothalamic electrode implantation disrupts lithium chloride-based generalized aversion to sodium chloride by enhancing sodium appetite. *Physiological Psychology*, **11(1)**, 63–72.

Masotto, C. & Negro-Vilar, A. (1985). Inhibition of spontaneous or angiotensin II-stimulated water intake by atrial natriuretic factor. *Brain Research Bulletin*, **15**, 523–6.

Massi, M. & Epstein, A.N. (1987). The apparent dependence of salt appetite in the pigeon on endogenous angiotensin II. *Physiology and Behavior*, **41**, 155–62.

Massi, M. & Epstein, A.N. (1989). Suppression of salt intake in the rat by neurokinin A: comparison with the effect of kassinin. *Regulatory Peptides*, **24**, 233–44.

Massi, M., & Epstein, A.N. (1990). Angiotensin/aldosterone synergy governs the salt appetite of the pigeon. *Appetite*, **14**, 181–92.

Massi, M., Gentili, L., Perfumi, M., deCaro, G., & Schulkin, J. (1990). Inhibition of salt appetite in the rat following injection of tachykinins into the medial amygdala. *Brain Research*, **513**, 1–7

Massi, M., Perfumi, M., deCaro, G. & Epstein, A.N. (1988). Inhibitory effect of kassinin on salt intake induced by different natriorexigenic treatments in the rat. *Brain Research*, **440**, 232–42.

Masson, D.B. & Fitts, D.A. (1989). Subfornical organ connectivity and drinking to captopril or carbachol in rats. *Behavioral Neuroscience*, **4**, 873–80.

Mattes, R.D. (1984). Salt taste and hypertension: A critical review of the literature. *Journal of Chronic Disease*, 1–14.

Mayer, J. (1969). Hypertension, salt intake, and the infant. *Postgraduate Medicine, Clinical Nutrition*, **45(1)**, 229–30.

Meikle, A.D.S. & Kaufman, S. (1988). Stretch-induced reduction in atrial content of natriuretic factor is locally mediated. *American Journal of Physiology*, R284–8.

Melby, J.C., Dale, S.L. & Wilson, T.E. (1971). 18-hydroxy-deoxycorticosterone in human hypertension. *Circulation Research*, Supplement H to XXVIII and XXXIX, 11–143–11–152.

Mendelsohn, F.A.O., Allen, A.M., Clevers, J., Denton, D.A., Tarjan, E. & McKinley, M.J. (1988). Localization of angiotensin II receptor binding in rabbit brain by *in vitro* autoradiography. *Journal of Comparative Neurology*, **270**, 372–84.

Menting, J., Morgan, T., Barrett, G. & DiNicolantonio, R. (1987). The effect of DOCA and 9a-Fluorocortisone on renal renin content and production. *Clinical and Experimental Pharmacology and Physiology*, **14**, 259–62.

Mercer, P.F., Mogenson, G.J. & Paquette, S.Y.M. (1978). Sodium intake following destruction of the anterior hypothalamus in the rat. *Canadian Journal of Physiology and Pharmacology*, **56**, 252–9.

Meyer, W.J. & Nichols, N.R. (1981). Mineralocorticoid binding in cultured smooth muscle cells and fibroblasts from rat aorta. *Journal of Steroid Biochemistry*, **14**, 1157–68.

Michell, A.R. (1975). Changes of sodium appetite during the estrous cycle of sheep. *Brief Communication. Physiology and Behavior*, **14**, 223–6.

Michell, A.R. (1978). Plasma potassium and sodium appetite; the effect of potassium infusion in sheep. *British Veterinary Journal*, **134**, 217–24.

Michell, A.R. (1979). Sodium transport and salt appetite: the effect of DPH on sodium preference and electrolyte balance in rats. *Chemical Senses and Flavour*, **4(3)**, 231–40.

Michell, A.R. (1986). The gut: the unobtrusive regulator of sodium balance. *Perspectives in Biology and Medicine*, **4**, 203–9.

Midkiff, E.E. & Bernstein, I.L. (1983). The influence of age and experience on salt preference of the rat. *Developmental Psychobiology*, **16(5)**, 385–94.

Midkiff, E.E., Fitts, D.A., Simpson, J.B. & Bernstein, I.L. (1985). Absence of sodium chloride preference in Fischer-344 rats. *American Journal of Physiology*, R438–42.

Miller, N.E. (1971a). *Selected Papers on Conflict, Displacement, Learned Drives and Theory.* Chicago: Aldine-Atherton Inc.

Miller, N.E. (1971b). *Selected Papers on Learning, Motivation and Their Physiological Mechanisms.* Chicago: Aldine-Atherton Inc.

Miller, W.L. (1988). Molecular biology of steroid hormone synthesis. *Endocrine Reviews*, **9(3)**, 295–312.

Mimran, A., Guiod, L. & Hollenberg, N.K. (1974). The role of angiotensin in the cardiovascular and renal response to salt restriction. *Kidney International*, **5**, 348–55.

Miselis, R.R. (1981). The efferent projections of the subfornical organ of the rat: A circumventricular organ within a neural network subserving water balance. *Brain Research*, **230**, 1–23.

Mistlberger, R.E. & Rechtschaffen, A. (1985). Periodic water availability is not a potent zeitgeber for entrainment of circadian locomotor rhythms in rats. *Physiology and Behavior*, **34**, 17–22.

Mizukami, S., Nishizuka, M. & Arai, Y. (1983). Sexual difference in nuclear volume and its ontogeny in the rat amygdala. *Experimental Neurology*, **79**, 569–75.

Moe, K.E. (1985). The ontogeny of salt preference in rats. *Developmental Psychobiology*, **19**, 185–96.

Moe, K.E. (1986). The ontogeny of salt intake in rats. In: A.N. Epstein, M. Massi and G. de Caro (eds), *The Physiology of Thirst and Sodium Appetite*: New York: Plenum Press.

Moe, K.E., Weiss, M.L. & Epstein, A.N. (1984). Sodium appetite during captopril blockade of endogenous angiotensin II formation. *American Journal of Physiology*, R356–65.

Montiel, M., Jimenez, E., Narvaez, J.A. & Morell, M. (1983). Renin–angiotensin–aldosterone system in hyper- and hypothyroid rats during sodium depletion. *Endocrine Research Communications*, **9(3&4)**, 249–60.

Mook, D.G. (1963). Oral and postingestional determinants of the intake of various solutions in rats with esophageal fistulas. *Journal of Comparative and Physiological Psychology*, **56(4)**, 645, 659.

Mook, D.G. (1969). Some determinants of preference and aversion in the rat. In: *Neural Regulation of Food and Water Intake*. P.J. Morgane (ed.). *Annals of the New York Academy of Sciences*, **157(2)**, 1158–75.

Morrison, G.R. (1969). The relative effectiveness of salt stimuli for the rat. *Canadian Journal of Psychology*, **23(1)**, 34–40.

Morrison, G.R. (1970). Detectability and preference for sodium chloride and sodium carbonate. *Physiology and Behavior*, **8**, 25–8.

Morrison, G.R. (1971). Effects of formalin-induced Na deficiency on CaCl and KCl acceptability. *Psychonomic Science*, **25(3)**, 167–8.

Morrison, G.R. & Young, J.C. (1972). Taste control over sodium intake in sodium deficient rats. *Physiology and Behavior*, **8**, 29–32.

Mouw, D.R., Vander, A.J. & Wagner, J. (1978). Effects of prenatal and early postnatal sodium-deprivation on subsequent adult thirst and salt preference in rats. *American Journal of Physiology*, **234(1)**, F59–63.

Mulrow, P.J., Takagi, M., Atarashi, K. & Franco-Saenz, R. (1987). Inhibition of aldosterone secretion by atrial natriuretic peptide. *Annals New York Academy of Science*, **512**, 438–48.

Munaro, N. & Chiaraviglio, E. (1981). Hypothalamic levels and utilization of noradrenaline and 5-hydroxytryptamine in the sodium-depleted rat. *Pharmacology Biochemistry and Behavior*, **15**, 1–5.

Murphy, H.M. & Brown, T.S. (1970). Effects of hippocampal lesions on simple and preferential consummatory behavior in the rat. *Journal of Comparative and Physiological Psychology*, **72(3)**, 404–15.

Mutter, J., Lemoine, J., Tsang, B. & Kucharczyk, J. (1984). Central angiotensin-induced water intake and salt appetite in the pig. *Brain Research*, **322**, 374–7.

Nachman, M. (1962). Taste preferences for sodium salts by adrenalectomized rats. *Journal of Comparative and Physiological Psychology*, **55(6)**, 1124–9.

Nachman, M. (1963a). Learned aversion to the taste of lithium chloride and generalization to other salts. *Journal of Comparative and Physiological Psychology*, **56(2)**, 343–9.

Nachman, M. (1963b). Taste preferences for lithium chloride by adrenalectomized rats. *American Journal of Physiology*, **205(2)**, 219–21.

Nachman, M. & Ashe, J.H. (1974). Effects of basolateral amygdala lesions on neophobia, learned taste aversions, and sodium appetite in rats. *Journal of Comparative and Physiological Psychology*, **87(4)**, 622–43.

Nachman, M. and Pfaffmann, C. (1963). Gustatory nerve discharge in normal and sodium-deficient rats. *Journal of Comparative and Physiological Psychology*, **56(6)**, 1007–11.

Nachman, M. & Valentino, D.A. (1966). Roles of taste and postingestional factors in the satiation of sodium appetite in rats. *Journal of Comparative and Physiological Psychology*, **62(2)**, 280–3.

Nakamura, M., Katsuura, G., Nakao, K. & Imura, H. (1985). Antidipsogenic action of a human atrial natriuretic polypeptide administered intracerebroventricularly in rats. *Neuroscience Letters*, **58**, 1–6.

Nardi, J.D., Stoner, E., Martin, K., Balfe, J.W., Jose, P.A. & New, M.I. (1987). New findings in apparent mineralocorticoid excess. *Clinical Endocrinology*, **27**, 49–62.

Nauta, W.J.H. (1961). Fiber degeneration following lesions of the amygdaloid complex in the monkey. *Journal of Anatomy*, **95(4)**, 515–31.

Nauta, W.J.H. and Domesick V.B. (1979). The anatomy of the extrapryamidal system. In: *Dopaminergic Ergot Derivatives and Motor Function*. K. Fuxe and D.B. Caine (eds), New York: Pergamon Press.

Needleman, P. & Greenwald, J.E. (1986). Atriopeptin: a cardiac hormone intimately invoved in fluid, electrolyte, and blood-pressure homeostasis. *New England Journal of Medicine*, **314(13)**, 828–9.

Nelson, D.O. & Boulant, J.A. (1981). Altered CNS neuroanatomical organization of spontaneously hypertensive (SHR) rats. *Brain Research*, **226**, 119–30.

Nicolaidis, S. (1969). Early systemic responses to orogastric stimulation in the regulation of food and water balance: functional and electrophysiological data. *Annals of the New York Academy of Science*, **157**, 1176–203.

Nicolaidis, S. (1977). Sensory–neuroendocrine reflexes and their anticipatory and optimizing role on metabolism. *The Chemical Senses and Nutrition*, Chap. 6, New York: Academic press.

Nicolaidis, S. & Jeulin, A-C (1986). Integrative rostrodiencephalic neurons and integrative peptides in hydromineral regulation. In: *Emotion Neuronal and Chemical Control* (ed.) Y. Oomure, J.A.A.P.

Nicolaidis, S., Galaverna, O. & Meltzer, C.G. (1990). Extracellular dehydration during pregnancy increases salt intake of offspring. *American Journal of Physiology*, **27**, R281–3.

Nishizuka, M. & Arai, Y. (1981). Sexual dimorphism in synaptic organization in the amygdala and its dependence on neonatal hormone environment. *Brain Research*, **212**, 31–8.

Nitabach, M.N. Schwartz, G., Spector, A.C. & Grill, N.J. (1988). The anterior lingual receptor field is responsible for recognition of sodium in rats. *Neuroscience Abstracts*.

Nitabach, M.N., Schulkin, J. & Epstein, A.N. (1989). The medial amygdala is part of a mineralocorticoid sensitive circuit controlling NaCl intake in the rat. *Behavioral Brain Research*, **35**, 127–34.

Norgren, R. (1970). Gustatory responses in the hypothalamus. *Brain Research*, **21**, 63–77.

Norgren, R. (1976). Taste pathways to hypothalamus and amygdala. *Journal of Comparative Neurology*, **166**, 17–30.

Norgren, R. (1984). Central neural mechanisms of taste. In: *Handbook of Physiology – The Nervous System* III. Sensory Processes, pt. 2. 1. Darian-Smith, vol. 4 J.M. Brookhart and V.B. Mountcastle (eds), Bethesda, MD: American Physiological Society, 1087–128.

Norgren, R. & Leonard, C.M. (1971). Taste pathways in rat brainstem. *Science*, **173**, 1136–9.

Norgren, R. & Pfaffmann, C. (1975). The pontine taste area in the rat. *Brain Research*, **91**, 99–117.

Nottebohm, F. & Arnold, A.P. (1976). Sexual dimorphism in vocal control areas of the songbird brain. *Science*, **194**, 211–13.

Novakova, A. & Stevenson, J.A.F. (1971). Effect of posterior hypothalamic lesions on renal function in the rat. *Canadian Journal of Physiology and Pharmacology*. **49(11)**, 941–50.

Nowlis, G.H. (1977). From reflex to representation: taste-elicited tongue movements in the human newborn. In: *Taste and Development* J.M. Weiffenbach (ed.), Bethesda, Maryland: US Department of Health, Education and Welfare.

O'Kelly, L.I., Falk, J.L. & Flint, D. (1958). Water regulation in the rat: 1. Gastrointestinal exchange rates of water and sodium chloride in thirsty animals. *Journal of Comparative and Physiological Psychology*, **51(1)**, 16–21.

Ohman, L.E. & Johnson, A.K. (1986). Lesions in lateral parabrachial nucleus enhance drinking to angiotensin II and isoproterenol. *American Journal of Physiology*, R504–9.

Olton, D.S. & Markowska, A.L. (1989). The effects of preoperative experience upon postoperative performance of rats with lesions of the hippocampal system. In: *Preoperative Events: Their Effects on Behavior Following Brain Damage*. J. Schulkin (ed.) New Jersey: Erlbaum Press.

Osborne, T.B. & Mendel, L.B. (1918). The choice between adequate and inadequate diets, as made by rats. *Journal of Biological Chemistry*, **35**, 19–27.

Osborne, P.G., Blair-West, J.R., Denton, D.A., McBurnie, M., Tarjan, E., Williams, R.M., & Weisinger, R.R. (1990). Decreased cerebral sodium concentration and sodium appetite in BALB/c mice. *American Journal of Physiology*, **28**, R741–4.

Ottersen, O.P. (1982). Connections of the amygdala of the rat. IV: Corticoamygdaloid and intraamygdaloid connections as studied with axonal transport of horseradish peroxidase. *Journal of Comparative Neurology*, **205**, 30–48.

Palkovits, M., Eskay, R.L. & Antoni, F.A. (1987). Atrial natriuretic peptide in the medial eminence is of paraventricular nucleus origin. *Neuroendocrinology*, **46**, 542–4.

Palmieri, G.M.A. & Taleisnik, S. (1969). Intake of NaCl solution in rats with diabetes insipidus. *Journal of Comparative and Physiological Psychology*, **68(1)**, 38–44.

Pangborn, R.M. & Pecord, S.D. (1982). Taste perception of sodium chloride in relation to dietary intake of salt. *American Journal of Clinical Nutrition*, **35**, 510–20.

Pasley, J.N., Koike, T.I. & Nelson, H.L. (1977). Absence of sodium appetite in cyclophosphamide and DOCA treated house mice. *Pharmacology Biochemistry and Behavior*, **6**, 265–7.

Paulus, R.A., Eng, R. & Schulkin, J. (1984). Preoperative latent place learning preserves salt appetite following damage to the central gustatory system. *Behavioral Neuroscience*, **98(1)**, 146–51.

Paxinos, G., Emson, P.C. & Cuello, A.C. (1978). Substance P projections to the entopeduncular nucleus, the medial preoptic area and the lateral septum. *Neuroscience Letters*, **7**, 133–6.

Peterson, J.G. (1942). Salt feeding habits of the house finch. *The Condor*, **44**, 73.

Pettinger, W.A., Marchelle, M. & Angusto, L. (1971). Renin suppression by DOC and NaCl in the rat. *American Journal of Physiology*, **221**, 1071–4.

Pfaff, D.W. (1980). *Estrogens and Brain Function: Neural Analysis of a Hormone-controlled Mammalian Reproductive Behavior*. New York: Springer-Verlag.

Pfaffmann, C. & Bare, J.K. (1950). Gustatory nerve discharges in normal and adrenalectomized rats. *Journal of Comparative and Physiological Psychology*, **43(4)**, 320–4.

Pfaffmann, C. (1952). Taste preference and aversion following lingual denervation. *Journal of Comparative and Physiological Psychology*, **45(5)**, 393–400.

Pfaffmann, C. (1955). Gustatory nerve impulses in rat, cat and rabbit. *Journal of Neurophysiology*, **18**, 429–40.

Pfaffmann, C. (1960). The pleasures of sensation. *Psychological Review*, **67**, 253–68.

Pfaffmann, C. (1967). The sense of taste. In: *Handbook of Physiology*, section. Alimentary Canal Volume I: Food and Water Intake C.F. Code (ed.), Washington, D.C.: American Physiological Society.

Pfaffmann, C., Frank, M. & Norgren, R. (1979). Neural mechanisms and behavioral aspects of taste. *Annual Review Psychology*, **30**, 283–325.

Pfaffmann, C., Norgren, R. & Grill, H.J. (1977). Sensory affect and motivation. In: *Tonic Functions of Sensory Systems*. B.M. Wenzel and H.P. Zeigler (eds) vol. 290, New York: New York Academy of Sciences.

Phoenix, C., Goy, R. Gerall, A. & Young, W. (1959). Organizing action of prenatally administered testosterone propionate on the tissues mediating mating behavior in the female guinea pig. *Endocrinology*, **65**, 369–82.

Pike, R.L. & Yao, C. (1971). Increased sodium chloride appetite during pregnancy in the rat. *Journal of Nutrition*, **101**, 169–76.

Plunkett, L.M. & Saavedra, J.M. (1985). Increased angiotensin II binding affinity in the nucleus tractus solitarius of spontaneously hypertensive rats. *Proceedings of the National Academy of Sciences, USA*, **82**, 7721–4.

Plunkett, L.M., Shigematsu, K., Kurihara, M. & Saavedra, J.M. (1987). Localization of angiotensin II receptors along the anteroventral third ventricle area of the rat brain. *Brain Research*, **405**, 205–12.

Porter, G.A., Bogoroch, R. & Edelman, I.S. (1964). On the mechanism of action of aldosterone on sodium transport: The role of RNA synthesis. *Proceedings of the National Academy of Sciences, USA*, **52(6)**, 1326–33.

Porter, J.J. & Relinger, H. (1972). Daily sodium intake as a function of time of measurement and formalin injection volume and concentration. *Psychonomic Science*, **26(5)**, 276–8.

Powley, T.L. (1977). The ventralmedial hypothalamic syndrome, satiety and a cephalic phase hypothesis. *Psychological Review*, **84**, 89–126.

Pressley, L. & Funder, J.W. (1975). Glucocorticoid and mineralocorticoid receptors in gut mucosa. *Endocrinology*, **97(3)**, 588–96.

Price J.L. (1981). Toward a consistent terminology for the amygdaloid complex. *The Amygdaloid Complex, INSERM Symposium* No. 20, Y. Ben-Ari (ed.) North-Holland Biomedical Press: Elsevier.

Quartermain, D. & Wolf, G. (1967). Drive properties of mineralocorticoid-induced sodium appetite. *Physiology and Behavior*, **2**, 261–3.

Quartermain, D. Wolf, G. & Keselica, J. (1969). Relation between medial hypothalamic damage and impairments in regulation of sodium intake. *Physiology and Behavior*, **4**, 101–3.

Quirk, S.J., Gannell, J.E. & Funder, J.W. (1983). Aldosterone-binding sites in pregnant and lactating rat mammary glands. *Endocrinology*, **8**, 1812–17.

Ragan, C., Ferrebee, J.W., Phyfe, P., Atchley, D.W. & Loeb, R.F. (1940). A syndrome of polydipsia and polyuria induced in normal animals by desoxycorticosterone acetate. *American Journal of Physiology*, **2**, 73–8.

Raisman, G. & Field, P.M. (1973). Sexual dimorphism in the neuropil of the preoptic area of the rat and its dependence on neonatal androgen. *Brain Research*, **54**, 1–29.

Ramsay, D.J. & Ganong, W.F. (1977). CNS regulation of salt and water intake. *Hospital Practice*, **10**, 63–69.

Ramsay, D.J. & Reid, I.A. (1981). Salt appetite in dogs. *Neuroscience Abstracts*.

Rapp, J.P. & Dahl, L.K. (1971). 18-Hydroxy-deoxycorticosterone secretion in experimental hypertension in rats. *Circulation Research*, **XXXVIII** and **XXIX**. 11–153, 11–159.

Reagan, L.P. Xyehai, Y., Mir, R. DePalo, L.R. & Fluharty, S.J. (1990). Upregulation of angiotensin II receptors by *in vitro* differentiation of murine N1E-115 neuroblastoma cells. *American Society for Pharmacology and Experimental Therapeutics*, **38**, 878–86.

Reis, D.J. (1981). The brain and arterial hypertension: Evidence for a neural-imbalance hypothesis. *American Journal of Physiology*, **13**, 87–102.

Relman, A.S. and Schwartz, W.B. (1952). The effect of DOCA on electrolyte balance in normal man and its relation to sodium chloride intake. *Yale Journal of Biology and Medicine*, **10**, 540–58.

Rescorla, R.A. (1981). Simultaneous Associations, In: *Predictability, Correlation and Contiguity*, P. Harzem and M.D. Zeiler (eds), John Wiley & Sons Ltd.

Reul, J.M.H.M., van den Bosch, F.R. & deKloet, E.R. (1987). Differential response of type I and type II corticosteroid receptors to changes in plasma steroid level and circadian rhythmicity. *Neuroendocrinology*, **45**, 407–12.

Ricardo, J.A. & Koh, E.T. (1978). Anatomical evidence of direct projections from the nucleus of the solitary tract to the hypothalamus, amygdala, and other forebrain structures in the rat. *Brain Research*, **153**, 1–26.

Rice, K.K. & Richter, C.P. (1943). Increased sodium chloride and water intake of normal rats treated with desoxycorticosterone acetate. *Endocrinology*, **33**, 106–15.

Richter, C.P. (1936). Increased salt appetite in adrenalectomized rats. *American Journal of Physiology*, **115**, 155–61.

Richter, C.P. (1939). Salt taste thresholds of normal and adrenalectomized rats. *Endocrinology*, **24**, 367–71.

Richter, C.P. (1941). Sodium chloride and dextrose appetite of untreated and treated adrenalectomized rats. *Endocrinology*, **29**, 115–25.

Richter, C.P. (1943). Total self regulatory functions in animals and human beings. *Harvey Lecture Series*, **38**, 63–103.

Richter, C.P. (1956). Salt appetite of mammals: Its dependence on instinct and metabolism. *L'instinct dans le comportement des animaux et de l'homme.* (pp. 577–629). Paris: Masson.

Richter, C.P. (1965). *Biological Clocks in Medicine and Psychiatry*. Springfield: Charles C. Thomas.

Richter, C.P. (1976). In: *The Psychobiology of Curt Richter*. E.M. Blass (ed.), Baltimore: New York.

Richter, C.P. & Barelare, B. Jr (1938). Nutritional requirements of pregnant and lactating rats studied by the self-selection method. *Endocrinology*, **23**, 15–24.

Richter, C.P. & Eckert, J.F. (1938). Mineral metabolism of adrenalectomized rats studied by the appetite method. *Endocrinology*, **21**, 214–24.

Richter, C.P. & MacLean, A. (1939). Salt taste threshold of humans. *American Journal of Physiology*, **126(1)**, 1–6.

Rodgers, W.L. (1967). Specificity of specific hungers. *Journal of Comparative and Physiological Psychology*, **64(1)**, 49–58.

Rogers, R.C. & Hermann, G.E. (1983). Central connections of the hepatic branch of the vagus nerve: a horseradish peroxidase histochemical study. *Journal of the Autonomic Nervous System*, **7**, 165–74.

Rogers, R.C., Kahrilas, P.J. & Hermann, G.E. (1984). Projection of the hepatic branch of the splanchnic nerve to the brainstem of the rat. *Journal of the Autonomic Nervous System*, **II**, 223–5.

Rogers, R.C., Novin, D. & Butcher, L.L. (1979). Electrophysiological and neuroanatomical studies of hepatic portal osmo- and sodium-receptive afferent projections within the brain. *Journal of the Autonomic Nervous System*, **1**, 183–202.

Rolls, B.J. & Rolls, E.T. (1982). Thirst. *Problems in the Behavioural Sciences*. Cambridge: Cambridge University Press.

Rosenwasser, A.M. & Adler, N.T. (1986). Structure and function in circadian timing systems: evidence for multiple coupled circadian oscillators. *Neuroscience and Biobehavioral Reviews*, **10**, 431–48.

Rosenwasser, A.M., Schulkin, J. & Adler, N.T. (1985). Circadian wheel-running activity of rats under schedules of limited daily access to salt. *Chronobiology International*, **2(2)**, 115–19.

Rosenwasser, A.M., Schulkin, J. & Adler, N.M. (1988). Anticipatory appetitive behavior of adrenalectomized rats under circadian salt-access schedules. *Animal Learning and Behavior*, **16(3)**, 324–9.

Rowland, N.E. (1986). Comparative physiological psychology of feeding and salt appetite in rodents. *Nutrition and Behavior*, **3**, 27–41.

Rowland, N.E. & Fregly, M.J. (1988a). Sodium appetite: Species and strain differences in the induction of sodium appetite: role of renin–angiotensin–aldosterone system. *Appetite*, **11**, 143–78.

Rowland, N.E. & Fregly, M.J. (1988b). Induction of an appetite for sodium in rats that show no spontaneous preference for sodium chloride solution – The Fischer 344 strain. *Behavioral Neuroscience*, **102(6)**, 961–8.

Rowland, N.E. & Fregly, M.J. (1988c). Characteristics of thirst and sodium appetite in mice (*Mus musculus*). *Behavioral Neuroscience*, **102(6)**, 969–74.

Rowland, N.E., Bellush, L.L. & Fregly, M.J. (1985). Nycthemeral rhythms and sodium chloride appetite in rats. *American Journal of Physiology*, R375–8.

Rozin, P. (1976). The selection of food by rats, humans, and other animals. In: *Advances in the Study of Behavior* J.S. Rosenblatt, R.A. Hinde, E. Shaw & C. Beer (eds) vol. 6 New York: Academic Press.

Rozin, P. & Schulkin, J. (1990). Food selection. In: *Handbook of Behavioral Neurobiology*. E.M. Stricker (eds), New York: Plenum Press.

Rozin, P. & Zellner, D.A. (1985). The role of Pavlovian conditioning in the acquisition of food likes and dislikes. *Annals of the New York Academy of Sciences*, **443**, 189–202.

Ruger, J. & Schulkin, J. (1980). Preoperative sodium appetite experience and hypothalamic lesions in rats. *Journal of Comparative and Physiological Psychology*, **94(5)**, 914–20.

Sakai, R.R. (1986). The hormones of renal sodium conservation act synergistically to arouse sodium appetite in the rat. In: *The Physiology of Thirst and Sodium Appetite*, de Caro, Epstein and Massi (eds). pp. 425–30. New York: Plenum Press.

Sakai, R.R. & Epstein, A.N. (1990a). Peripheral angiotensin II is not the cause of sodium appetite in the rat. *Appetite*, **15**, 161–70.

Sakai, R.R. & Epstein, A.N. (1990b). The dependence of adrenalectomy-induced sodium appetite on the action of angiotensin II in the brain of the rat. *Behavioral Neuroscience*, **104**, 167–76.

Sakai, R.R., Fine, W.B., Epstein, A.N. & Frankmann, S.P. (1987). Salt appetite is enhanced by one prior episode of sodium depletion in the rat. *Behavioral Neuroscience*, **101(5)**, 724–31.

Sakai, R.R., Frankmann, S.P., Fine, W.B. & Epstein, A.N. (1989). Prior episodes of sodium depletion increase the need-free sodium intake of the rat. *Behavioral Neuroscience*, **103(1)**, 186–92.

Sakai, R.R., Nicolaidis, S. & Epstein, A.N. (1986). Salt appetite is suppressed by interference with angiotensin II and aldosterone. *American Journal of Physiology*, **251**, R762–8.

Sakai, R.R., Witcher, J.A., Adler, N.T. & Epstein, A.N. (1988). Sexual dimorphism of need-free salt intake in the rat. *Eastern Psychological Association*, 59.

Samson, W.K. (1985). Atrial natriuretic factor inhibits dehydration and hemorrhage-induced vasopressin release. *Neuroendocrinology*, **40**, 277–9.

Saper, C.B. (1982a). Convergence of autonomic and limbic connections in the insular cortex of the rat. *Journal of Comparative Neurology*, **210**, 163–73.

Saper, C.B. (1982b). Reciprocal parabrachial–cortical connections in the rat. *Brain Research*, **242**, 33–40.

Saper, C.B. (1983). Afferent connections of the median preoptic nucleus in the rat: anatomical evidence for a cardiovascular integrative mechanism in the ante-roventral third ventricular (AV3V) region. *Brain Research*, **288**, 21–31.

Saper, C.B., Reis, D.J., & Joh, T. (1983). Medullary catecholamine inputs to the anteroventral third ventricular cardiovascular regulatory region in the rat. *Neuroscience Letters*, **42**, 285–91.

Scalia, F. & Winans, S.S. (1975). The differential projections of the olfactory bulb and accessory olfactory bulb in mammals. *Journal of Comparative Neurology*, **161**, 31–56.

Schaller, G.B. (1963). *The Mountain Gorilla*. Chicago: University of Chicago Press.

Schiffman, S.A. (1980). Contribution of the anion to the taste quality of sodium salt. In: *Biological and Behavioral Aspects of Salt Intake*. M. Kare, M. Fregly, and B. Bernard (eds) New York: Academic Press.

Schiffman, S.S., Lockhead, E. & Maes, F.W. (1987). Amiloride reduces the taste intensity of Na$^+$ and Li$^+$ salts and sweeteners. *Proceedings of the National Academy of Sciences, USA*, **80**, 6136–40.

Schiffman, S.S., Simon, S.A., Gill, J.M. & Beeker, T.G. (1986). Bretylium tosylate enhances salt taste. *Physiology and Behavior*, **36**, 1129–37.

Schmidt, M. (1973). Influences of hepatic portal receptors on hypothalamic feeding and satiety centers. *American Journal of Physiology*, **225(5)**, 1089–95.

Schmidt-Nielsen, K. (1964, 1979). *Desert Animals, Physiological Problems of Heat and Water*. New York: Dover Publications.

Schneirla, T.C. (1959, 1972). An evolutionary and developmental theory of biphasic process underlying approach and withdrawal. In: *Selected Writings of T.C. Schneirla*. L.R. Aronson, E. Tobach, J.S. Rosenblatt and D.S. Lehrman (eds) San Francisco: W.H. Freeman & Co.

Schulkin, J. (1978). Mineralocorticoids, dietary conditions, and sodium appetite. *Behavioral Biology*, **23**, 197–205.

Schulkin, J. (1982). Behavior of sodium deficient rats: The search for a salty taste. *Journal of Comparative and Physiological Psychology*, **96**, 628–34.

Schulkin, J. (1986). The evolution and expression of salt appetite. In: *The Physiology of Thirst and Sodium Appetite*. G. de Caro, A.N. Epstein and M. Massi (eds), New York: Plenum Publishing Corp.

Schulkin, J. (1989). In honor of a great inquirer: Curt Richter. *Psychobiology*, **17**, 113–14.

Schulkin, J. (1988). The effects of preoperative ingestive events on feeding and drinking behavior following brain damage. *Psychobiology*, **16**, 185–95.

Schulkin, J. & Fluharty, S.J. (1985). Further studies on salt appetite following lateral hypothalamic lesions: effects of preoperative alimentary experiences. *Behavioral Neurosciences*, **99(5)**, 929–35.

Schulkin, J. & Grill, H.J. (1980). Compensatory ingestion in the decorticate rat. *International Conference of Food and Fluid Intake*, Warsaw, Poland.

Schulkin, J. & Grill, H.J. (1984). Performance versus competence deficits in insulin induced feeding in the decorticate rat. *Neuroscience Abstract*.

Schulkin, J. & Ruger, J. (1980). Relation between lateral hypothalamic damage and impairment of sodium appetite: evidence of subcortical mass action. *Behavioral and Neural Biology*, **30**, 90–6.

Schulkin, J., Arnell, P. & Stellar, E. (1985a). Running to the taste of salt in mineralocorticoid-treated rats. *Hormones and Behavior*, **19**, 413–25.

Schulkin, J., Eng. R. & Miselis, R.R. (1983). The effects of disconnecting the subfornical organ on behavioral and physiological responses to alterations of body sodium. *Brain Research*, **263**, 351–5.

Schulkin, J., Flynn, F.W., Grill, H.J. & Norgren, R. (1985b). Central gustatory lesions II: effects on salt appetite and taste aversion learning. *Neuroscience Abstract*, **11**, 1259.

Schulkin, J., Liebman, D., Ehrman, R.N., Norton, N.W. & Ternes, J.W. (1984). Salt hunger in the rhesus monkey. *Behavioral Neuroscience*. **98(4)**, 753–6.

Schulkin, J., Tordoff, M.G. & Friedman, M.I. (1987). Contribution of taste and postabsorptive factors in the satiation of salt hunger. *Federation of American Societies of Experimental Biology*.

Schulkin, J., Angulo, J., Sakai, R. and McEwen, B.S. (1989b). Expression of salt hunger in hypophysectomized rats. *Neuroendocrinology*, **50**, 447–53.

Schulkin, J., Marini, J. & Epstein, A.N. (1989*a*). A role for the medial region of the amygdala in mineralocorticoid induced salt hunger. *Behavioral Neuroscience*, **103**, 178–185.

Schwaber, J.S., Kapp, B.S., Higgins, G.A. & Rapp, P.R. (1982). Amygdaloid and basal forebrain direct connections with the nucleus of the solitary trace and the dorsal motor nucleus. *Journal of Neuroscience*, **2(10)**, 1424–38.

Schwartz, G.J. & Grill, H.J. (1984). Relationships between taste reactivity and intake in the neurologically intact rat. *Chemical Senses*, **9(3)**, 249–72.

Schwartz, G.J. & Grill, H.J. (1985). Comparing taste-elicited behaviors in adult and neonatal rats. *Appetite*, **6**, 373–86.

Sclafani, A. & Nissenbaum, J.W. (1985). On the role of the mouth and gut in the control of saccharin and sugar intake: a reexamination of the sham-feeding preparation. *Brain Research Bulletin*, **14**, 569–76.

Scott, E.M., Verney, E.L. & Morissey, P.D. (1950). Self selection of diet XI, Appetites for calcium, magnesium and calcium. *Journal of Nutrition*, **41**, 187–202.

Scott, T.R. & Chang, F-C, T. (1984). The state of gustatory neural coding. *Chemical Senses*, **8(3)**, 297–314.

Scott, T.R. and Mark, G.P. (1986). Feeding and taste. *Progress in Neurobiology*, **27**, 293–317.

Selye, H. (1952). *The Story of the Adaptation Syndrome*, Montreal: ACTA Inc.

Sernia, C., Sinton, L., Thomas, W.G. & Pascoe, W. (1985). Liver angiotensin II receptors in the rat: binding properties and regulation by dietary Na^+ and angiotensin II. *Journal of Endocrinology*, **106**, 103–11.

Shapiro, M.D. & Linas, S.L. (1985). Sodium chloride pica secondary to iron-deficiency anemia. *American Journal of Kidney Diseases*, **V(1)**, 67–8.

Shapiro, R.E. & Miselis, R.R. (1985). The central neural connections of the area postrema of the rat. *Journal of Comparative Neurology*, **234**, 344–64.

Shepherd, R. & Farleigh, C.A. (1986). Preferences, attitudes and personality as determinants of salt intake. *Human Nutrition: Applied Nutrition*, **40A**, 195–208.

Shiono, K. & Sokabe, H. (1976). Renin–angiotensin system in spontaneously hypertensive rats. *American Journal of Physiology*, **231(4)**, 1295–9.

Shulkes, A.A., Covelli, M.D., Denton, D.A. & Nelson, J.F. (1972). Hormonal factors influencing salt appetite in lactation. *Australian Journal of Experimental Biology and Medical Science*, **7**, 819–26.

Shrager, E.E. & Johnson, A.K. (1980). Anteroventral third ventricle (AV3V) region ablation: chronic elevations of plasma renin concentration. *Brain Research*, **190**, 554–8.

Sills, M.A., Nguyen, K.Q. & Jacobowitz, D.M. (1985). Increased in heart rate and blood pressure produced by microinjections of atrial natriuretic factor into the Av3V region of rat brain. *Peptides*, **6**, 1037–42.

Simerly, R.B. & Swanson, L.W. (1986). The organization of neural inputs to the medial preoptic nucleus of the rat. *Journal of Comparative Neurology*, **246**, 312–42.

Simpson, J.B. (1975). Subfornical organ involvement in angiotensin-induced drinking. In: *Control Mechanisms of Drinking* G. Peters, J.T. Fitzsimons & L. Peters-Haefeli (eds) New York: Springer-Verlag.

Skofitsch, G., Jacobowitz, D.M., Eskay, R.L. & Zamir, N. (1985). Distribution of atrial natriuretic factor-like immunoreactive neurons in the rat brain. *Neuroscience*, **16(4)**, 917–48.

Sly, J. and Bell, F.R. (1979). Experimental analysis of the seeking behavior observed in ruminants when they are sodium deficient. *Physiology and Behavior*, **22**, 499–505.

Smith, D.F. & Stricker, E.M. (1969). The influence of need on the rat's preference for dilute NaCl solutions. *Physiology and Behavior*, **4**, 407–10.

Smith, D.F., Stricker, E.M. & Morrison, G.R. (1969). NaCl solution acceptability by sodium-deficient rats. *Physiology and Behavior*, **4**, 239–43.

Smith, D.F., Bealer, S.L. & Van Buskirk, R.L. (1978). Adaptation and recovery of the rat chorda tympani response to NaCl. *Physiology and Behavior*, **20**, 629–36.

Snyder, Jr R.L. & Sutin, J. (1961). Effect of lesions of the medulla oblongata on electrolyte and water metabolism in the rat. *Experimental Neurology*, **4**, 424–35.

Sonnenberg, H., Milojevic, S., Chong, C.K. & Veress, A.T. (1983). Atrial natriuretic factor: reduced cardiac content in spontaneously hypertensive rats. *Hypertension*, **5(5)**, 672–5.

Soulairac, A. (1969). The adrenergic and cholinergic control of food and water intake. *Annals of The New York Academy of Sciences*, **157(2)**, 934–61.

Spector, A.C., Schwartz, G.S. & Grill, H.J. (1990). Chemospecific deficits in taste detection following selective gustatory deafferentation. *American Journal of Physiology*, in press.

Spigel, I.M., Ellis, K.R. & Kaiser, Y.E. (1967). Electrolyte balance and drinking in the fresh water turtle. *Journal of Comparative and Physiological Psychology*, **64**, 313–17.

Spyer, K.M. (1981). Neural organisation and control of the baroreceptor reflux. *Review in Physiology and Biochemistry and Pharmacology*, **88**, 23–121.

Steinberg, J. & Bindra, D. (1962). Effects of pregnancy and salt-intake on genital licking. *Journal of Comparative and Physiological Psychology*, **55**, 103–6.

Steiner, J. (1977). Facial expressions of the neonate infant indicating the hedonics of food-related chemical stimuli. In: *Taste and Development: The Genesis of Sweet Preference* J.M. Weiffenbach (ed.), Bethesda, MD: US Department of Health, Education, and Welfare.

Stellar, E. (1954). The physiology of motivation. *Psychological Review*, **61**, 5–22.

Stellar, E. (1980). Brain mechanisms and hedonic processes. *Acta Neurobiology Experimentia*, **40**, 313–24.

Stellar, E. (1987). The internal environment and appetitive measures of taste function in the rat. In: *Perspectives in Chemoreception and Behavior*. R.F. Chapman, E.A. Bernays, and J.G. Stoffolano, Jr (eds), New York: Springer-Verlag.

Stellar, E., Hyman, R. & Samet, S. (1954). Gastric factors controlling water and salt solution drinking. *Journal of Comparative and Physiological Psychology*, **47(3)**, 220–6.

Stellar, J.R. & Stellar, E. (1985). *The Neurobiology of Motivation and Reward*. New York: Springer-Verlag.

Stellar, J.R. Brooks, F.H. & Mills, L.E. (1979). Approach–withdrawal analysis of the effects of hypothalamic stimulation and lesions in rats. *Journal of Comparative and Physiological Psychology*, **93**, 446–66.

Sterc, J., Novakova, V. & Golda, V. (1975). Effects of dorsal, mediobasal and laterobasal septal lesions in the rat: water and sodium chloride intake. *Experimental Neurology*, **48**, 175–88.

Stevenson, J.A.F., Welt, L.G. & Orloff, J. (1950). Abnormalities of water and electrolyte metabolism in rats with hypothalamic lesions. *American Journal of Physiology*, **161**, 35–9.

Stewart, P.M., Wallace, A.M., Valentino, R., Burt, D., Shackleton, C.H.L. & Edwards, C.R.W. (1987). Mineralocorticoid activity of liquorice: II – beta-hydroxysteroid dehydrogenase deficiency comes of age. *Lancet*, **ii**, 821–3.

Stoppini, L. & Baertschi, A.J. (1984). Activation of portal–hepatic osmoreceptors in rats: role of calcium, acetylcholine and cyclic AMP. *Journal of the Autonomic Nervous System*, **11**, 297–308.

Streeten, D.H.P. & Rapoport, A. (1963). Existence of a slowly exchangeable pool of body sodium in normal subjects and its diminution in patients with primary aldosteronism. *Journal of Clinical Endocrinology and Metabolism*, **XXIII**, 928–37.

Stricker, E.M. (1966). Extracellular fluid volume and thirst. *American Journal of Physiology*, **211(1)**, 232–8.

Stricker, E.M. (1969). Osmoregulation and volume regulation in rats: inhibition of hypovolemic thirst by water. *American Journal of Physiology*, **217(1)**, 98–105.

Stricker, E.M. (1981). Thirst and sodium appetite after colloid treatment in rats. *Journal of Comparative and Physiological Psychology*. **95(1)**, 1–24.

Stricker, E.M. (1983). Thirst and sodium appetite after colloid treatment in rats: role of the renin–angiotensin–aldosterone system. *Behavioral Neuroscience*, **97(5)**, 725–37.

Stricker, E.M. & Jalowiec, J.E. (1970). Restoration of intravascular fluid volume following acute hypovolemia in rats. *American Journal of Physiology*, **218(1)**, 191–6.

Stricker, E.M. & Sterritt, G.M. (1967). Osmoregulation in the newly hatched domestic chick. *Physiology and Behavior*, **2**, 117–19.

Stricker, E.M. & Verbalis, J.G. (1987). Central inhibitory control of sodium appetite in rats: correlation with pituitary oxytocin secretion. *Behavioral Neuroscience*, **101(4)**, 560–7.

Stricker, E.M. & Wilson, N. (1970). Salt-seeking behavior in rats following acute sodium deficiency. *Journal of Comparative and Physiological Psychology*, **72(3)**, 416–20.

Stricker, E.M. & Wolf, G. (1966). Blood volume and tonicity in relation to sodium appetite. *Journal of Comparative and Physiological Psychology*, **62(2)**, 275–9.

Stricker, E.M. & Wolf, G. (1969). Behavioral control of intravascular fluid volume: thirst and sodium appetite. *Annals of New York Academy of Sciences*, **157(2)**, 553–68.

Stricker, E.M. & Zigmond, M.J. (1974). Effects on homeostasis of intraventricular injections of 6-hydroxydopamine in rats. *Journal of Comparative and Physiological Psychology*, **86(6)**, 973–94.

Stricker, E.M., Hosutt, J.A. & Verbalis, J.G. (1987). Neurohypophyseal secretaion in hypovolemic rats: inverse relation to sodium appetite. *American Journal of Physiology*, R889–96.

Stricker, E.M., Vagnucci, A.H., McDonald, R.H. Jr & Leenen, F.H. (1979). Renin and aldosterone secretions during hypovolemia in rats: relation to NaCl intake. *American Journal of Physiology*, R45–51.

Stumpf, W.E. & Sar, M. (1979). Glucocorticoid and mineralocorticoid hormone target sites in the brain: Autoradiographic studies with corticosterone, aldosterone and dexamethasone. In: *Interaction Within the Brain Pituitary – Adrenocortical System*,

M.T. Jones, M.F. Dallman, B. Gillham, and S. Chattopadhyay, (eds), New York: Academic Press.

Swanson, L.W. & Mogenson, G.J. (1981). Neural mechanisms for the functional coupling of autonomic endocrine and somatomotor responses in adaptive behavior. *Brain Research Reviews*, **3**, 1–34.

Tanaka, J. & Seto, K. (1988). Neurons in the lateral hypothalamic area and zona incerta with ascending projections to the subfornical organ area in the rat. *Brain Research*, **456**, 397–400.

Tang, M. & Falk, J.L. (1979). Temporary peritoneal sequestration of NaCl and persistent NaCl appetite. *Physiology and Behavior*, **22**, 595–7.

Tarjan, E., Denton, D.A. & Weisinger, R.S. (1988). Atrial natriuretic peptide inhibits water and sodium intake in rabbits. *Regulatory Peptides*, **23**, 63–75.

Tarjan, E., Denton, D.A. McKinley, M.H., Nelson, J.F. and Weisinger, R.R. (1984). What makes wild rabbirts drink? *Journal of Physiology*, **79**, 46–71.

Tarttelin, M.F. & Gorski, R.A. (1971). Variations in food and water intake in the normal and acyclic female rat. *Physiology and Behavior*, **7**, 847–52.

Thompson, C.I. & Epstein, A.N. (1991). Salt appetite in rat pups: ontogeny of angiotensin II – aldosterone synergy *American Journal of Physiology*, **260**, R421–9.

Thrasher, T.N. & Fregly, M.J. (1980). Factors affecting salivary sodium concentration, NaCl intake, and preference threshold and their interrelationship. In: *Biological and Behavioral Aspects of Salt Intake*, (eds) M.R. Kare, M.J. Fregly, and R.A. Bernard, New York: Pergamon Press.

Thrasher, T.N., Keil, L.C. & Ramsay, D.J. (1982). Lesions of the organum vasculosum of the lamina terminalis (OVLT) attenuate osmotically-induced drinking and vasopressin secretion in the dog. *Endocrinology*, **12**, 1837–9.

Thunhorst, R.L., Fitts, D.A. & Simpson, J.B. (1987). Separation of captopril effects on salt and water intake by subfornical organ lesions. *American Journal of Physiology*, **252**, R409–18.

Titlebaum, L.F., Falk, J.L. & Mayer, J. (1960). Altered acceptance and rejection of NaCl in rats with diabetes insipidus. *American Journal of Physiology*, **199(1)**, 22–4.

Toates, F. (1986). *Motivational Systems*. New York: Cambridge University Press.

Tolman, E.C. (1949). *Purposive Behavior in Animals and Men*. Berkeley: University of California Press.

Tordoff, M.G. & Friedman, M.I. (1986). Hepatic portal glucose infusions decrease food intake and increase food preference. *American Journal of Physiology*, **251** (*Regulatory Integrative Comparative Physiology*, **20**), R192–6.

Tordoff, M.G., Schulkin, J. & Friedman, M.K. (1986). Hepatic contribution to satiation of salt appetite in rats. *American Journal of Physiology*, R1095–102.

Tordoff, M.G., Schulkin, J. & Friedman, M.I. (1987). Further evidence for hepatic control of salt intake in rats. *American Journal of Physiology*, R444–9.

Tordoff, M.G., Ulrich, P.M. & Schulkin, J. (1990). Calcium deprivation increases salt intake. *American Journal of Physiology*, R411–19.

Toth, E., Stelfox, J. & Kaufman, S. (1987). Cardiac control of salt appetite. *American Journal of Physiology*, R925–9.

Travers, J.B. & Smith, D.V. (1979). Gustatory sensitivities in neurons of the hamster nucleus tractus solitarius. *Sensory Processes* **3**, 1–26.

Travers, J.B., Grill, H.J. & Norgren, R. (1987). The effects of glossopharyngeal and chorda tympani nerve cuts on the ingestion and rejection of sapid stimuli: an electromyographic analysis in the rat. *Behavioral Brain Research*, **25**, 233–46.

Trent, A.M. & Kalat, J.W. (1977). Lake of effect of specific sodium hunger on learned aversions to sodium and sucrose. *Animal Learning and Behavior*, **5(3)**, 243–6.

Turner, B.H., Mishkin, M. & Knapp, M. (1980). Organization of the amygdalopetal projections from modality – specific cortical association areas in the monkey. *Journal of Comparative Neurology*, **191**, 515–43.

Uysal, S., Schulkin, J. & Hyde, T.M. (1985). The effects of AP-cmNTS lesions or subdiaphragmatic vagotomy on salt appetite. *Neuroscience Abstracts.*

Van der Kooy, D., Koda, L.Y., McGinty, J.F., Gerfen, C.R. & Bloom, F.E. (1984). The organization of projections from the cortex, amygdala, and hypothalamus to the nucleus of the solitary tract in rat. *Journal of Comparative Neurology*, **224**, 1–24.

Vance, W.B. (1965). Observations on the role of salivary secretions in the regulation of food and fluid intake in the white rat. In: *Psychological Monographs* G.A. Kimble (ed.), American Psychological Association, **79(5)**, No. 598.

VanHemel, P.E. (1976). Ingestion of potassium chloride and soium chloride solutions by albino and hooded rats. *Behavioral Biology*, **17**, 519–27.

Vari, R.C., Freeman, R.H., Davis, J.O., Villarreal, D. & Verburg, K.M. (1986). Effect of synthetic atrial natriuretic factor on aldosterone secretion in the rat. *American Journal of Physiology*, R48–52.

Verburg, K.M., Freeman, R.H., Davis, J.O., Villarreal & Vari, R.C. (1986). Control of atrial natriuretic factor release in conscious dogs. *American Journal of Physiology*, R947–56.

Veress, A.T. & Sonnenberg, H. (1984). Right atrial appendectomy reduces the renal response to acute hypervolemia in the rat. *American Journal of Physiology*, R610–13.

Vijande, M., Costales, M. & Marin, B. (1977). Sex difference in polyethylenglycol-induced thirst. *Experientia*, **34**, 742–3.

Vijande, M., Costales, M., Schiaffini, O. & Marin, B. (1978). Angiotensin-induced drinking: Sexual differences. *Pharmacology Biochemistry and Behavior*, Brief Communication, **8**, 753–5.

Vilar, A.N., Gentil, C.G. & Covian, M.R. (1967). Alterations in sodium chloride and water intake after septal lesions in the rat. *Physiology and Behavior*, **2**, 167–70.

Vinson, G.P., Whitehouse, B.J., Goddard, C. & Sibley, C.P. (1979). Comparative and evolutionary aspects of aldosterone secretion and zona glomerulosa function. *Journal of Endocrinology*, 5P–24P.

Vivas, L. & Chiaraviglio, E. (1987). Effect of agents which alter the Na transport on the sodium appetite in rats. *Brain Research Bulletin*, **19**, 679–85.

Wade, G.N. & Zucker, I. (1969a). Hormonal and developmental influences on rat saccharin preferences. *Journal of Comparative and Physiological Psychology*, **69(2)**, 291–300.

Wade, G.N. & Zucker, I. (1969b). Taste preferences of female rats: modification by neonatal hormones, food deprivation and prior experience. *Physiology and Behavior*, **4**, 935–43.

Walsh, L.L. & Grossman, S.P. (1977). Electrolytic lesions and knife cuts in the region of the zona incerta impair sodium appetite. *Physiology and Behavior*, **18**, 587–96.

Weeks, H.P. Jr & Kirkpatrick, C.M. (1976). Adaptations of white-tailed deer to naturally occurring sodium deficiencies. *Journal of Wildlife Management*, **40**, 610–25.

Weiner, I.H., & Stellar, E. (1951). Salt preference of the rat determined by a single-stimulus method. *Journal of Comparative and Physiological Psychology*, **44(4)**, 394–401.

Weisinger, R.S. & Woods, S.C. (1971). Aldsterone-elicited sodium appetite. *Endocrinology*, **89**, 538–44.

Weisinger, R.S. (1975). Conditioned and pseudoconditioned thirst and sodium appetite. In: *Control Mechanisms of Drinking*, G. Peters. J.T. Fitzsimons & L. Peters-Haefeli (eds), New York: Springer-Verlag.

Weisinger, R.S., Denton, D.A. & McKinley, M.J. (1983). Self-administered intravenous infusion of hypertonic solutions and sodium appetite of sheep. *Behavioral Neuroscience*, **97(3)**, 433–44.

Weisinger, R.S., Coghlan, J.P., Denton, D.A., Fan, J. Hatzikostas, S., McKinley, M.J., Nelson, J.F. & Scoggins, B.A. (1980). ACTH induced Na appetite in sheep. *American Journal of Physiology* **239**, 645–50.

Weisinger, R., Considine, P. Denton, D., Di Nicolantonio, R., McKinley, M.J., Muller, A.F. & Tarjan, E. (1982). Role of sodium concentration of the cerebrospinal fluid in the salt appetite of sheep. *American Journal of Physiology*, **242** (*Regulatory Integrative Comparative Physiology*), **11**, R51–63.

Weisinger, R., Denton, D., Di Nicolantonio, R., McKinley, M.J., Muller, A.F. & Tarjan, E. (1987a). Role of angiotensin in sodium appetite of sodium-deplete sheep. *American Journal of Physiology*, **253** (*Regulatory Integrative Comparative Physiology*, **22**). R482–8.

Weisinger, R.S., Denton, D.A., McKinley, M.J. & Nelson, J.F. (1978). ACTH induced sodium appetite in the rat. *Pharmacology Biochemistry and Behavior*, **8**, 339–42.

Weisinger, R.S., Denton, D.A., McKinley, M.J. & Nelson, J.F. (1985a). Dehydration-induced sodium appetite in rats. *Physiology and Behavior*, **34**, 45–50.

Weisinger, R., Denton, D., McKinley, M.J., Muller, A.F. & Tarjan, E. (1985b). Cerebrospinal fluid sodium concentration and salt appetite. *Brain Research*, **326**, 95–105.

Weisinger, R., Denton, D., McKinley, M.J., Muller, A.F. and Tarjan, E. (1986). Angiotensin and Na appetite of sheep. *American Journal of Physiology*, **251**, (*Regulatory Integrative Comparative Physiology*), **20**, R690–9.

Weisinger, R.S., Blair-West, J.R., Denton, D.A., McBurnie, M., Ong., F., Tarjan, E. & Williams, R.M. (1990). Effect of angiotensin-converting enzyme inhibitor on salt appetite and thirst of BALB/c mice. *American Journal of Physiology*, **28**, R736–40.

Weisinger, R.S., Denton, D.A., McKinley, M.J., Osborne, P.G. & Tarjan, E. (1987b). Decrease of brain extracellular fluid (Na) and its interaction with other factors influencing sodium appetite in sheep. *Brain Research*, **420**, 135–43.

Weisinger, R.S., Denton, D.A., Di Nicolantonio, R., McKinley, M.J., Oldfield, B. & Osborne, P.G. (1990). Subfornical organ lesion decreases sodium appetite in the sodium-depleted rat. *Brain Research*, **526**, 23–30.

Weisinger, R.S., Woods, S.C. & Skorupski, J.D. (1970). Sodium deficiency and latent learning. *Psychonomic Science*, **19(5)**, 307–8.

Weiss, M.L., Moe, K.E. & Epstein, A.N. (1986). Interference with central actions of angiotensin II. *American Journal of Physiology*. R250–9.

Welch, W.J., Ott, C.E., Lorenz, J.N. & Kotchen, T.A. (1987). Control of renin release by dietary NaCl in the rat. *American Journal of Physiology*, F1052–7.

Wilkins, L. & Richter, C.P. (1940). A great craving for salt by a child with cortico-adrenal insufficiency. *JAMA Clinical Notes, Suggestions and New Instruments*. In: *Food Selections*, 866–8.

Will, P.C., Cortright, R.N., DeLisle, R.C., Douglas, J.G. & Hopfer, U. (1985). Regulation of amiloride-sensitive electrogenic sodium transport in the rat colon by steroid hormones. *American Journal of Physiology*, G124–31.

Will, P.C., Lebowitz, J.L. & Hopfer, U. (1980). Induction of amiloride-sensitive sodium transport in the rat colon by mineralocorticoids. *American Journal of Physiology*, F261–8.

Wilson, C.S. & Wong, R. (1975). Aldactazide-induced consummatory and operant responding to sodium by rats. *American Journal of Psychology*, **88(3)**, 377–84.

Wilson, K.M., Sumners, C., Hathaway, S. & Fregly, M.J. (1986). Mineralocorticoids modulate central angiotensin II receptors in rats. *Brain Research*, **382**, 87–96.

Wirsig, C.R. & Grill, H.J. (1982). Contribution of the rat's neocortex to ingestive control: I. Latent learning for the taste of sodium chloride. *Journal of Comparative and Physiological Psychology*, **96(4)**, 615–27.

Wolf, G. (1964*a*). Effect of dorsolateral hypothalamic lesions on sodium appetite elicited by desoxycorticosterone and by acute hyponatremia. *Journal of Comparative and Physiological Psychology*, **58(3)**, 396–402.

Wolf, G. (1964*b*). Sodium appetite elicited by aldosterone. *Psychonomic Science*, **1**, 211–12.

Wolf, G. (1965). Effect of deoxycorticosterone on sodium appetite of intact and adrenalectomized rats. *American Journal of Physiology*, **208(6)**, 1281–5.

Wolf, G. (1967). Hypothalamic regulation of sodium intake: relations to preoptic and tegmental function. *American Journal of Physiology*. **213(6)**, 1433–8.

Wolf, G. (1968*a*). Projections of thalamic and cortical gustatory areas in the rat. *Journal of Comparative Neurology*, **132(4)**, 519–30.

Wolf, G. (1968*b*). Thalamic and tegmental mechanisms for sodium intake: Anatomical and functional relations to lateral hypothalamus. *Physiology and Behavior*, **3**, 997–1002.

Wolf, G. (1969*a*). Effects of a mineralocorticoid antagonist on sodium appetite. *VIII International Congress of Nutrition*, Prague.

Wolf, G. (1969*b*). Innate mechanisms for regulation of sodium intake. In: *Olfaction and Taste*. C. Pfaffman (ed.), New York: Rockefeller University Press.

Wolf, G. (1971). Neural mechanisms for sodium appetite: Hypothalamus positive–hypothalamofugal pathways negative. *Physiology and Behavior*, **6**, 381–9.

Wolf, G. (1982). Refined salt appetite methodology for rats demonstrated by assessing sex differences. *Journal of Comparative and Physiological Psychology*, **96(6)**, 1016–21.

Wolf, G. & Handal, P.J. (1966). Aldosterone-induced sodium appetite: Dose-response and specificity. *Endocrinology*, **78(6)**, 1120–4.

Wolf, G. & Lawrence, G.H. (1963). Saline preference curve for mice: lack of relationship to pigmentation. *Nature, London*, **200(4910)**, 1025–6.

Wolf, G. & Quartermain, D. (1966). Sodium chloride intake of desoxycorticosterone-treated and of sodium-deficient rats as a function of saline concentration. *Journal of Comparative and Physiological Psychology*, **61(2)**, 288–91.

Wolf, G. & Quartermain, D. (1967). Sodium chloride intake of adrenalectomized rats with lateral hypothalamic lesions. *American Journal of Physiology*, **212(1)**, 113–16.

Wolf, G. & Schulkin, J. (1980). Brain lesions and sodium appetite: An approach to the neurological analysis of homeostatic behavior. In: *Biological and behavioral aspects of salt intake*. M. Kare (ed.), New York: Academic Press.

Wolf, G. & Stricker, E.M. (1967). Sodium appetite elicited by hypovolemia in adrenalectomized rats: Reevaluation of the 'reservoir' hypothesis. *Journal of Comparative and Physiological Psychology*, **63(2)**, 252–7.

Wolf, G., Dahl, L.K. & Miller, N.E. (1965). Voluntary sodium chloride intake of two strains of rats with opposite genetic susceptibility to experimental hypertension. *Experimental Biology and Medicine*, **120**, 301–5.

Wolf, G., Dicara, L.V. & Braun, J.J. (1970). Sodium appetite in rats after neocortical ablation. *Physiology and Behavior*, 1265–9.

Wolf, G., McGovern, J.F. & Di Cara, L.V. (1974). Sodium appetite: some conceptual and methodologic aspects of a model drive system. *Behavioral Biology*, **10**, 27–42.

Wolf, G., Schulkin, J. & Fluharty, S.J. (1983). Recovery of salt appetite after lateral hypothalamic lesions: Effects of preoperative salt drive and salt intake experiences. *Behavioral Neuroscience*, **97(3)**, 506–11.

Wolf, G., Schulkin, J. & Simpson, P.E. (1984). Multiple factors in the satiation of salt appetite. *Behavioral Neuroscience*, **98(4)**, 661–73.

Wong, R. (1977). Saline intake in gerbils (*Meriones unguiculatus*). *Physiological Psychology*, **5(2)**, 225–9.

Wong, R. (1981). Maintenance diet and the effects of furosemide on hamsters. *American Journal of Psychology*, **94(2)**, 339–54.

Wong, R. & Jones, W. (1978). Effects of aldactazide and DOCA injections on saline preference in gerbils (*Meriones unguiculatus*). *Behavioral Biology*, **23**, 460–8.

Wong, R. & Kraintz, L. (1977). Desalivation and saline ingestion in rats. *Behavioral Biology*, **19**, 130–4.

Wong, R. & Whiteside, C.B.C. (1974). Sodium preference induced by angiotensin in the rat. *Pharmacological Research Communications*, **6(3)**, 307–10.

Woodbury, D.M. (1956). Effect of acute hyponatremia on distribution of water and electrolytes in various tissues of the rat. *American Journal of Physiology*, **185(2)**, 281–6.

Woodbury, D.M. & Koch, A. (1957). Effects of aldosterone and desoxycorticosterone on tissue electrolytes. *Proceedings of the Society of Experimental Biology and Medicine*, 720–3.

Woods, S.C., Vasseli, J.R. Kaestner, E., Szakmary, G.A., Milbern, P. & Vitello, M.V. (1977). Conditioned insulin secretion and meal feeding in rats. *Journal of Comparative and Physiological Psychology*, **91**, 128–33.

Woodside, B. & Millelire, L. (1987). Self-selection of calcium during pregnancy and lactation in rats. *Physiology and Behavior*, **39**, 291–5.

Wright, J.W. & Donlon, K. (1979). Inhibition of starvation-induced body dehydration by saline consumption in rats and gerbils. *Behavioral and Neural Biology*, **25**, 535–44.

Wright, J.W., Reynolds, T.J. & Kenny, J.T. (1975). Plasma hyperosmolality at the onset of drinking during starvation induced hypovolemia. *Physiology and Behavior*, **17**, 651–7.

Wyss, J.M. & Donovan, M.K. (1984). A direct projection from the kidney to the brainstem. *Brain Research*, **298**, 130–4.

Yagil, Y., Koreen, R. & Krakoff, L.R. (1986). Role of mineralocorticoids and glucocorticoids in blood pressure regulation in normotensive rats. *American Journal of Physiology*, H1354–9.

Yensen, R. (1958). Influence of salt deficiency on taste sensitivity in human subjects. *Nature, London*, **181**, 1472–4.

Yensen, R. (1959). Some factors affecting taste sensivity in man. II: Depletion of body salt. *Quarterly Journal of Experimental Psychology*, **XI(4)**, 230–8.

Yongue, B.G. & Myers, M.M. (1988). Cosegregation analysis of salt appetite and blood pressure in genetically hypertensive and normotensive rats. *Clinical and Experimental Hypertension-Theory and Practice*, **A10(2)**, 323–43.

Yongue, B.G. & Roy, E.J. (1987). Endogenous aldosterone and corticosterone in brain cell nuclei of adrenal-intact rats: regional distribution and effects of physiological variations in serum steroids. *Brain Research*, **436**, 49–61.

Yoshii, K., Kiyomoto, Y. & Kurihara, K. (1986). Taste receptor mechanism of salts in frog and rat. *Comparative Biochemical Physiology*, **85A(3)**, 501–7.

Young, P.T. (1949). Studies of food preference, appetite and dietary habit. *Comparative Psychology Monographs*, **19**, 1–74.

Young, P.T. (1952). The role of hedonic processes in the organization of behavior. *Psychological Review*. **59**, 249–62.

Young, P.T. (1959). The role of affective processes in learning and motivation. *Psychological Review*, **66**, 104–25.

Young, P.T. & Falk, J.L. (1956). The relative acceptability of sodium chloride solutions as a function of concentration and water need. *Journal of Comparative and Physiological Psychology*, 569–75.

Young, P.T., Falk, J.L. & Kappauf, W.F. (1956). Running activity and preference as related to concentration of sodium–chloride solutions. *Journal of Comparative and Physiological Psychology*, **49**, 569–75.

Zhang, D.M., Stellar, E. & Epstein, A.N. (1984). Together intracranial angiotensin and systemic mineralocorticoid produce avidity for salt in the rat. *Physiology and Behavior*, **32**, 677–81.

Zicha, J., Panek, M., Salatova, J. & Krecek, J. (1972). The epiphysis and the effect of neonatal administration of testosterone propionate on salt intake in rat males. *Physiological Bohemoslov*. **21**, 453–4.

Zimmerman, M.B., Blaine, E.H. & Stricker, E.M. (1981). Water intake in hypovolemic sheep: effects of crushing the left atrial appendage. *Science*, **211**, 489–91.

Zolovick, A.J., Avrith, D. & Jalowiec, J.E. (1980). Reversible colchicine-induced disruption of amygdaloid function in sodium appetite. *Brain Research Bulletin*, **5**, 35–9.

Zucker, I. (1969). Hormonal determinants of sex differences in saccharin preference, food intake and body weight. *Physiology and Behavior*, **4**, 595–602.

Name index

Adam, W.R.   82, 83
Adler, N.T.   17, 19, 52
Aguilera, G.   88
Ahern, G.   16, 17
Aldrich, E.C.   7
Andersson, B.   104
Angullo, J.A.   96, 97, 112
Antunes-Rodriquez, J.   41
Arnell, P.   22
Arnold, A.P.   2, 3, 15, 31
Arriza, J.L.   113
Ashe, J.H.   121
Atarashi, K.   98
Avrith, D.B.   33

Baertschi, A.J.   101
Baldwin, B.A.   30, 135
Balment, R.J.   95
Barbella, Y.   97
Bare, J.K.   10, 59, 62, 75
Barelare, B. Jr   48, 49, 54
Bartoshuk, L.M.   59, 65, 70, 71, 81, 86
Bates, P.L.   52
Beach, F.A.   31
Bealer, S.L.   108, 118
Beauchamp, G.K.   30, 60, 70, 74, 108
Beaumont, K.   92
Beidler, L.M.   77
Beilharz, S.   104
Bell, F.R.   22, 30, 59, 78
Belvosky, G.E.   6
Ben-Ari, E.T.   99
Bergstrom, W.H.   102
Berl, T.   107
Bernard, C.   4, 5
Bernstein, I.L.   29, 64, 70, 77
Berridge, K.C.   4, 20, 21, 24, 25, 26, 57, 71, 73
Bertino, M.   12, 30, 60, 70, 74, 108, 129
Bianchi, C.   129
Bindra, D.   48

Bird,, E.   107
Birmingham, M.K.   113
Blair-West, J.R.   6, 30, 87, 91, 104
Blake, W.D.   76, 99
Blass, E.M.   79
Bliss, D.K.   52
Bolles, R.C.   11, 12
Borer, K.T.   76
Botkin, D.B.   6
Boudreau, J.C.   63, 77
Bour, T.C.   70
Brainard, J.B.   107
Braun, J.J.   121
Braun-Menendez, E.   31, 35, 38
Breedlove, S.M.   2, 3
Bregar, R.E.   15, 16
Brink, V.C.   6
Brinton, R.E.   93, 115
Brown, J.E.   51
Brown, T.S.   114, 135
Bryant, R.W.   33
Buggy, J.   33, 46, 47, 48, 52, 92, 118
Bunge, M.   103
Burnell, G.M.   107
Bursey, R.G.   107
Butkus, A.   108

Cabanac, M.   4, 9, 71
Camacho, A.   118
Cannon, W.B.   4, 5, 9, 60
Cantin, M.   97
Carpenter, J.A.   77, 78
Carr, W.J.   59
Casto, R.   108
Castren, E.   95
Catalanotto, F.A.   59, 108
Chang, F-C.T.   65, 69
Chao, H.M.   95, 113, 115
Chernigovsky, V.N.   76
Chiaraviglio E.   33, 77, 88, 90, 103, 104, 106, 118, 126, 132, 135

Chimoskey J.E.,   108
Claire, M.   92
Coirini, H.   93, 113, 115
Coleman, J.S.   7
Contreras, R.J.   4, 59, 62, 63, 64, 70, 78,
    107, 132, 134
Cooke, H.J.   103
Courtney, L.   70
Covey, E.   65
Cowan, I.   6, 85, 124
Cox, J.R.   121
Crabbe, J.   93
Craig, W.C.   8–9, 22
Cullen, J.W.   30, 82

Dahl, L.K.   108
Danielsen, J.   48
Darwin, C.   23
Davidson, R.J.   74
Dawborn, J.K.   82, 83
DeCaro, G.   131
Deems, R.,   79
DeKloet, E.R.   92
DeLuca, L.A. Jr   33, 118, 119, 128, 129
DeNicola, A.F.   92
Denton, D.A.   2, 4, 5, 6, 7, 11, 22, 23, 29,
    30, 31, 34, 35, 37, 38, 42, 43, 44, 49, 50,
    60, 70, 77, 78, 79, 81, 86, 87, 91, 95, 98,
    103, 104, 105, 108, 112, 118, 138, 140
Deolmos, J.D.   124, 125
Deschepper, C.F.   102
Desor, J.A.   70
Dethier, V.G.   58
Deutsch, J.A.   20, 75
Dickinson, A.   14, 15
Diezi, J.   91
DiLorenzo, P.M.   73
DiNicolantonio, R.   108
Dixon, J.S.   7
Donlon, K.   106
Donovan, M.K.   91
Douglas, J.   95, 103
Duncan, C.J.   77
Duval, D.   99

Edelman, I.S.   93
Edwards, G.L.   132
Ehrman, R.N.   53
Eisman, E.H.   14
Elfont, R.M.   90, 107
Epstein, A.N.   1, 2, 11, 27, 32, 33, 34, 35,
    37, 38, 39, 40, 54, 74, 90, 104, 119, 122,
    123, 124, 128, 129
Erickson, R.P.   65
Evans, R.M.   113
Evvard, J.M.   138

Falk, J.L.   2, 11, 26–7, 81, 76, 86
Farleigh, C.A.   70
Ferrario, C.M.   107–8
Findlay, A.L.R.   32, 46, 90
Fisher, A.E.   33, 92
Fitts, D.A.   30, 38, 40, 41, 54, 97, 99, 120,
    130, 132, 135
Fitzsimons, J.T.   30, 32, 33, 47, 88, 90, 99,
    101, 115
Fluharty, S.J.   2, 27, 28, 32, 33, 39, 90
Flynn, F.W.   66, 116, 117
Fonberg, E.   74
Forbes, G.B.   102
Foster, T.A.   107
Fox, N.A.   74
Frank, M.E.   62, 63, 64, 86
Fraser, J.D.   7
Fregly, M.J.   12, 28, 29, 30, 35, 37, 38, 46,
    52, 59, 77, 78, 95, 106
Friedman, M.   34, 77, 79, 80, 99, 134
Fudim, O.K.   20
Fuller, L.M.   33
Funder, J.W.   92, 93, 99

Galaverna, O.   119, 120, 129
Galef, B.J. Jr   21
Ganong, W.F.   87, 112
Garcia, J.   24
Garrison, J.C.   99
Genest, J.   97
Gentil, C.G.   77
Gibbs, J.   40, 99
Gibson, J.J.   6
Gibson, T.R.   129
Glaser, G.H.   107
Gomez-Sanchez, E.P.   108
Goodall, J.   7
Goy, R.W.   2, 54, 125
Grace, J.E.   77
Grace, S.A.   87
Green, D.M.   91, 92
Green, H.H.   138
Gregorova, I.   107
Grill, H.J.   15, 24, 29, 57, 64, 66, 68, 71, 73,
    76, 77, 116, 117, 130, 135
Grimsley, D.L.   12, 82
Grossman, L.   135
Grossman, S.P.   135

Hall, W.G.   79
Halpern, B.P.   57
Hamilton, W.J.   15
Hamlin, M.N.   37
Handal, P.J.   11, 12
Harrelson, A.   93
Harriman, A.E.   59, 77, 82

Harris, L.J.   9, 138
Hartzell, A.K.   16
Henkin, R.I.   60
Hennessy, C.J.   64
Herbert, D.   6, 85
Herman, T.S.   11
Hermann, G.E.   79
Herrick, C.J.   61, 69, 121
Herxheimer, A.   59
Hettinger, T.P.   64
Hilfehause, M.   95
Hill, D.L.   70, 71
Hoefer, R.   78, 104
Hoffman, R.A.   31
Horisberger, J-D.   91
Horky, K.   107
Hoshishima K.   78
Hubel, D.   70
Hull, C.L.   13
Hwang, B.H.   108
Hyde, T.M.   132–3

Ikonomov, O.C.   17
Israel, A.   97, 98

Jackson, J.H.   116
Jacobowitz, D.M.   130
Jacobs, K.M.   72
Jakinovich, W. Jr   86
Jalowiec, J.E.   81, 95, 101
Jerome, C.   101, 134
Jeulin, A-C.   80, 81
Johnson A.K.   112, 118, 132
Jones, A.D.   75
Jonklass, J   46, 47, 52

Katz, D.   138
Katzman, R.   102
Kaufman, S.   46, 98, 99
Kaunitz, H.   106
Kelly, T.M.   92
Kendrick, K.M.   135
Kevetter, G.A.   121
Khalil, K.A.   14
Kirkpatrick, C.M.   6
Kissileff, H.R.   78, 104
Koch, A.   93, 103
Koh, S.D.   59
Kosar, E.   68
Kosten, T.   62, 70, 101, 134
Kraintz, L.   60
Krecek, J.   2, 52, 53
Krettek, J.E.   124
Krieckhaus, E.E.   2, 8, 14, 16
Krozowski, Z.S.   93
Kucharczyk, J.   46, 47

Kurtz, T.W.   108–9
Kuta, C.C.   77
Kutscher, C.L.   106

Lan, N.C.   93
Lautt, W.W.   99
Lawrence, G.H.   78
Leaf, A.   107
LeMagnen, J.   82, 103
Leshem, M.   11, 33, 34, 54
Levy, C.J.   78
Levy, M.,   101
Lewis, M.   22, 85
Liebman, D.   53
Lin, K.K.   76, 99
Linas, S.L.   85
Lind, R.W.   111, 112, 118

Mackay, B.J.   46
MacLean, A.   58, 59
Macleod, R.B.   59
Maller, O.   70
Manaker, S.   32, 33, 90
Mann, H.   95
Marini, J.   54, 122, 123
Mark, G.P.   68, 71, 72
Markowska, A.L.   115
Marler, P.R.   15
Martin, J.R.   101, 134
Mason, D.B.   135
Masotto, C.   41
Massi, M.   35, 40, 41, 131
Masson, D.B.   120
Mattes, R.D.   109
Mayer, J.   108
McBurney, D.H.   51, 59
McBurnie, M.   131
McCance, R.A.   30, 60
McCleary, R.A.   76
McCoord, A.   102
McCutcheon, B.   78
McEwen, B.S.   2, 3, 44, 54, 93, 94, 96, 97,
    115, 121, 125
McKenzie, J.S.   77
McKinley, M.J.   104, 111, 112, 118
McMurray, T.M.   30
Meikle, A.D.S.   99
Mendel, L.B.   138
Michell, A.R.   103
Midkiff, E.E.   77
Millelire, L.   54
Miller, N.E.   1, 78
Miselis, R.R.   116, 132, 134
Mistlberger, R.E.   19
Mistretta, C.M.   70
Moe, K.E.   11, 13

Monkton, E.A.   99
Mook, D.G.   76
Moore-Gillon, M.J.   30, 99
Morris, R.C. Jr   108–9
Morrison, G.R.   14, 15
Mouw, D.R.   107
Mulrow, P.J.   98
Munaro, N.   106
Murphy, H.M.   114, 135
Mutter, J.   34
Myers, M.M.   108

Nachman, M.   4, 11, 62, 64, 65, 78, 81, 82,
   121
Nauta, W.J.H.   124
Negro-Vilar, A.   41
Nelson, D.H.   92
Nelson, J.F.   37, 49, 50
Nicholas, D.J.   14, 15
Nicholaidis, S.   70, 79, 80, 81, 107
Nissenbaum, J.W.   79
Nitabach, M.N.   63, 64, 121, 124, 125, 129
Norgren, R.   24, 66, 68, 69, 80–1, 116, 127
Norton, N.W.   53
Nottebohm, F.   15, 31
Novin, D.   101, 134
Nowlis G.H.   23, 61, 73

Ohman, L.E.   132
O'Kelly, L.I.   103
Olton, D.S.   115
Osborn, D.W.   86
Osborne, P.G.   104
Osborne, T.B.   138

Palkovits, M.   130
Palmieri, G.M.A.   107
Pangborn, R.M.
Paulus, R.A.   17, 18, 19
Pecord, S.D.   70
Perez-Guaita, M.F.   104, 118
Peterson, J.G.   7
Pettinger, W.A.   116
Pfaff, D.W.   3, 35, 38, 93
Pfaffmann, C.   3–4, 59, 60–1, 62, 69
Phillips, M.I.   108, 118
Phoenix, C.   31
Plunkett, L.M.   108
Powley, T.L.   80
Price, J.L.   124
Pringle, C.A.   7

Quartermain, D.   23, 37
Quirk, S.J.   49

Ramsay, D.J.   30, 40, 87
Rechtschaffen, A.   19

Reid, I.A.   30, 40
Reis, D.J.   108, 112, 132
Rescorla, R.A.   19, 20
Reul, J.M.H.M.   95
Rice, K.K.   35
Richter, C.P.   1, 2, 3, 4, 5, 9–10, 17, 35, 48,
   49, 54, 58, 59, 62, 63, 107
Ritter, R.C.   132
Robinson, P.F.   31
Rodgers, W.L.   12–13
Rogers, R.C.   134
Rolls, B.J.   104, 116, 132
Rolls, E.T.   104, 116, 132
Rosenwasser, A.M.   17, 19, 52
Rowland, N.E.   17, 21, 28, 29, 38, 78, 95
Roy, E.J.   93, 114
Rozin, P.   6, 8, 13, 19, 21, 26, 57, 86
Ruger, J.   27, 28

Saavedra, J.M.   95, 108
Sabine, J.R.   11
Sakai, R.   3, 27, 28, 29, 38, 52, 54, 90, 96,
   97, 116
Samson, W.K.   98
Saper, C.B.   118, 134
Sar, M.   113
Scalia, F.   121
Schaller, G.B.   7
Schmidt-Nielson, K.   60
Schulkin, J.   4, 6, 20, 21, 22, 23, 25, 26, 27,
   28, 30, 37, 41, 44, 53, 57, 64, 66, 67, 68,
   71, 79, 80, 81, 82, 83, 84, 85, 86, 95, 96,
   97, 100, 115, 117, 118, 119, 121, 122, 123,
   124, 128, 129, 130, 133, 135
Schwaber, J.S.   69
Schwartz, G.J.   64, 76
Schwartzbaum, J.S.   73
Sclafani, A.   79
Scott, T.R.   65, 68, 69, 71, 72
Selye, H.   42
Sernia, C.   99
Shakespeare, W.   15
Shapiro, M.D.   85
Shapiro, R.E.   132, 134
Shepherd, R.   70
Shiono, K.   107
Shulkes, A.A.   50, 51
Sills, M.A.   97
Simerly, R.B.   124
Simpson, J.B.   120
Simpson, P.E.   79
Skofitsch, G.   130
Sly, J.   22, 30
Smith, D.F.   11, 61, 69, 75
Smith, G.P.   101, 134
Snowden, C.T.   30
Sokabe, H.   107

Sokol, H.W.   107
Sonnenberg, H.   97, 108
Spector, A.C.   63, 64
Spigel, I.M.   58
Spyer, K.M.   108
Steilen, H.   106
Steinberg, J.   48
Steiner, J.   23
Stelfox, J.   98
Stellar, E.   1, 4, 9, 11, 22, 60, 71, 74, 75, 76, 87, 129
Stellar, J.R.   22, 69, 74, 76, 119, 128, 129
Sterrit, G.M.   21
Stetson, P.W.   132
Stevenson, J.A.F.   134
Stoppini, L.   101
Streeten, D.H.P.   102
Stricker, E.M.   5, 21, 32, 82, 87, 88, 90, 91, 93, 96, 98, 101, 102, 104, 106, 132
Stumf, W.E.   119

Taleisnik, S.   88, 106, 107, 132
Tang, M.   26–7
Tapper, D.N.   57
Tarjan, E.   41, 99, 104
Taylor, R.E. Jr   106
Teitelbaum, P.   59
Ternes, J.W.   53
Thompson, C.I.   34, 37, 38
Thrasher, T.N.   59, 80, 120
Thunhorst, R.L.   135
Titlebaum, L.F.   107
Tolman, E.C.   13, 16
Toma, R.B.   51
Tordoff, M.G.   12, 78, 79, 80, 82–3, 99, 100, 101, 134
Toth, E.   98, 99
Travers, J.B.   61, 69
Trent, A.M.   82
Turner, B.H.   124

Ulrich, P.M.   83
Uysal, S.   132, 133

Valentino, D.A.   78
Vallet, P.G.   101
Vance, W.B.   60
VanHemel, P.E.   74
Vari, R.C.   98

Verbalis, J.G.   132
Verburg, K.M.   97
Veress, A.T.   97
Vijande, M.   46
Vivas, L.   103

Wade, G.N.   46
Wallace, W.M.T.   102
Waters, I.W.   35, 37, 106
Watson, M.L.   107
Weeks, H.P. Jr, 6
Weiner, I.H.   74
Weisel, T.   70
Weisinger, R.S.   14, 22, 32, 34, 36, 40, 42–3, 90, 104, 106, 135
Weiss, M.L.   116
Wexler, M.J.   101
Whiteside, C.B.C.   33
Wilkins, L.   107
Will, P.C.   93
Williams, H.L.   59
Wilson, C.S.   22
Wilson, N.   82
Winans, S.S.   121
Wirsig, C.R.   15, 68
Wolf, G.   1, 2, 5, 12, 14, 16, 22, 23, 27, 35, 36, 37, 44, 45, 54, 55, 68, 78, 79, 87, 91, 93, 98, 101, 102, 104, 108, 110, 116, 135, 140
Wong, R.   22, 33, 38, 60
Woodbury, D.M.   59, 93, 103, 104
Woods, S.C.   36
Woodside, B.   54
Wright, J.W.   106
Wyss, J.M.   91

Yensen, R.   60
Yongue, B.G.   93, 108, 114
Yoshi, K.   58
Young, J.C.   14
Young, P.T.   4, 70–1, 73, 74, 76

Zellner, D.A.   19
Zhang, D.M.   33, 39
Zicha, J.   52
Zigmond, M.J.   106
Zimmerman, M.B.   98
Zolovick, A.J.   121
Zucker, I.   46

Subject index

acceptance system   57, 73
'acid-best' fibers   64
ACTH   42–5
ADH secretion   98
adrenal gland   91–4, 107
adrenalectomy   8, 9–10, 17, 33, 35, 36, 43,
    45, 58–9, 65, 76, 90, 92, 101, 106, 116
adrenergic agents   106
aldosterone   2, 6, 31, 35, 37, 43, 49, 88, 90,
    91, 95, 98, 99, 103–4, 108, 112–15, 121,
    123, 136, 136–7
'alliesthesia'   4
amiloride, 63–4, 93
amygdala   77, 135, 137
anatomy   33, 110–37
androgenization   47
angiotensinogen   95
angiotensinergic signal   96–7
angiotensin   2, 32–5, 38, 40, 46–7, 87–91,
    95, 99, 106, 107, 111, 116, 118–20, 124
animals, evolution of land animals   58
anterior ventral third ventricular region
    127, 136
anticipatory behavior   17
appetite
    central catecholamines   106
    gustation   60
    hormones   36
    physiology of the heart   99
    salt appetite   90, 101–2, 126, 135–6
appetitive phase   9, 22
area postrema, 132–4
atrial natriuretic factor   40–1, 92, 97–9, 108,
    129–31
atrial natriuretic peptide   97
atriopeptin   41
aversive responses   24
aversive substances   4
aversive/ingestive continuum   73
avidity for salt   91

basic salt taste category   65
behavior   5, 36, 71, 138
binding of angiotensin   108
binding of mineralocorticoids   92
biological clocks   17
birds   7, 28, 51
blood–brain barrier   111
body fluid regulation   97
body weight   34
bone   102, 138
brain
    aldosterone   112–15
    angiotensin system   111–12
    atrial natriuretic factor   40–1
    brain damage   17
    cholinergic action   101
    circuitry underlying sodium hunger
        135–7
    elevated salt intake   129
    females   125–6
    fluid ingestion   46–7
    gustation   60–1, 66–9
    hormone receptors   3
    medial amygdala   121
    physiological factors   90–1, 93
    potassium   104
    regions for natriorexigenic hormones
        116–18
    renin–angiotensin system   87–91
    salt appetite   126
    salt preference   76–7
brainstem   66, 132

calcium   83, 85, 101
captopril   116
carbachol   40, 46, 132
carnivores   7, 30, 40, 51, 77
catecholamines   106, 132
categories, gustation   65
central angiotensin induced thirst   47

189

central catecholamines   106
central gustatory sites   65–9
cephalic influences, gustation   80–1
cerebral spinal fluid   105
chalky taste   15
chickens   20–1
cholinergic action in the brain   101
chorda tympani   61–4, 74
circadian clock   19
circadian organization of renin–angiotensin–aldosterone system   95
circumventricular organs   111
citric acid   26
competent feeding and drinking   11
concentration of NaCl   74–5
conspecifics   21
consummatory phase   9, 24
contextual clues   15
contraceptives   52
cortex   68
corticosterone   88, 92–3, 115
cortisol   44
cranial nerves   60, 66, 73
cytosolic binding of aldosterone   115

dark phase cycle   17, 20, 95
decerebrated rats,   66–8, 71, 116
detection of salt   58–65
diabetes insipidus   107
dilute NaCl   76–7
diuresis   91–2
DOCA (deoxycorticosterone)   23, 32, 35, 38, 39, 44, 99, 108, 121, 127–8, 131
dogs   40, 80, 99
diopamine   77
drinkometer records   12

electron microscopy   110
electrophysiological signals   75
eledosin   131
elephants   85
endogenous biological clocks   17
endorphins   77
enzymes   87–8
Eskimo Indians   70
estrogen   46, 93
estrous cycles   46–7
evolution   31, 54, 58
expectancy, salt seeking behavior   22
extinction test   14, 16
extracellular fluid regulation   40–1, 46, 98, 101, 139
extraoral receptors   76

facial displays   23–6
females

appetite for salt   32
avidity for sodium   125–6
gustation   85
hormones   45–7, 54
male/female ratio   107
salt ingestion   125–6
salt seeking   2, 7, 138
serum levels of ALDO   91
fiber pathways   126
firing patterns of neurons   71
flavors   20
fluid ingestion   46–7
food   19, 106
forebrain   116
furosemide   13

gastric canulas   80
gastrointestinal tract   103
gender development   52–3
genital licking   48
gerbils   77
glial cells   102
glucocorticoid hormones   43–4
glucocorticoid receptor sites   93, 113
gonadal steroids   125
gonadectomy   54
gustation   3, 6, 17, 30, 57–86, 135, 138
gustatory nerve innervation   61
gustatory tuning   69

hamsters   78
heart,   40–1, 98–9, 108, 134
hedonic component to salt ingestion   3
hedonic perception of salt   23–8, 70–4
hepatic atrial afferents   134–5
hepatic infusion   79, 99–100, 134
hepatic sites   76
hepatic vagotomy   100–1
herbivores   7, 86, 104
hindbrain   134
hippocampus   113–14
homeostasis   4, 57, 87, 93, 107, 115, 118, 124, 131, 134, 136
hormones
   pathological conditions   107
   receptor localization   111–16
   regulation of salt intake   5, 31–56
humans
   detection of salts   58
   facial displays   23–4
   gustation   86
   salt seeking behavior   30
   salt taste development   70
hypertension   108, 134
hypertonic saline intake   36
hypophysectomized animals   95

hypothalamus   27, 28, 79–81, 134
hypothroidism   106
hypovolemic condition   75, 87, 98

identifying salty tastes   58–65
immunoreactive cell groups   112
ingestive behavior   3, 4, 25–6, 27
inhibition of salt intake   129–35
inland animals, salt seeking behavior   6
innate behavior   2
intracellular dehydration   46
intracerebral infusions of angiotensin   90
intragastric intubation   78
intraventricular infusions of hypertonic
     sodium   105

kassinin   131
kidneys   87–91, 107

lactation   46, 49–51
latent learning   14, 16, 21, 68
lateral hypothalamic damage   27
learning   2
   latent learning   68
   salt seeking behavior   9, 11, 13–21
leaves (plants)   86
lesion of brain   66–8, 118, 121–5, 135
licking patterns   64
ligand–dependent transcription factor   113
light–dark cycle   17, 20, 95
limbic connection   68
lithium chloride   82
litter size   49–50, 107
'little brains'   103
liver   79, 97, 99–101, 134
location of salt   15, 16
lordosis   3, 93
lungs   88

male/female ratio   107
mammals   6
manitol   105
markers for mineral licks   57
medial amygdala   121–9, 131
medial parabrachial region (PBN)   66
memory, salt seeking behavior   8, 16
microscopy, verse   110
mineral deficiency   81–6, 139
mineral deposits (licks)   57
mineral hunger   48
mineralocorticoids   31, 35–8, 38, 59, 91–4,
     102, 108, 11, 111–13, 121–4, 131
monkeys   52, 53
mothers' milk   11
motivation   1, 22–3, 71
mouse (mice)   78

narrow tuning   69
natriorexigenic conditions   17, 28, 32, 54,
     55, 103, 107
need-free preference   74–9
neocortex   15
neonates   11, 13, 33–4, 46, 47
nephrectomy   88
nerve fibres   62
nerves   60–5
neural circuit   139
neurohormones   131–2
neurons   71
nonsodium salts   9
noradrenaline   10
novel foods   13
nucleus medianus   120

olfactory projections   121
oral cavity   59, 71
oral contraceptives   52
oral stimulation   79
osmo or sodium receptor   104
osmotic regulation   132
ovarectomy   46
OVLT   120
oxytocin   49, 51, 132

palatability   4, 71, 73, 76
parabrachial gustatory region   135
perceptionof salt   70–2
peripheral nerves   60–5
phosphorus   138
physiology   87–109, 139
pica   85
pigeons   34, 38, 77
pigs   34, 138
pineal gland   52
pituitary   95–6
place learning   16, 19
place memory   57
plasma renin activity   88
plasma volume   98
'pleasures of sensations'   3–4
potassium   83, 86, 91, 103–6
precocious ingestive behavior   34
preference, salt preference   74–7
pregnancy   70
'prelimbic' area   77
preoperative salt ingestion   17
preoptic region   47, 97
prewiring (innate prewiring)   2
prolactin   46, 49, 50
protein   20, 93, 94
psychology and sodium   1
psychophysics   71
psychosis   107
purina chow   101

quinine 26

rabbits 37, 42, 44, 49, 78
rats 1, 8–21, 22, 29, 34, 37, 38, 42–3, 44,
    47, 48, 59, 62–3, 70, 74, 76, 79, 82–4,
    86, 90, 95, 98, 102, 103, 104, 107, 108,
    112, 115, 116, 120, 129, 130, 132–3, 136
receptors
    angiotensin 111
    mineralocorticoids 92
    receptor proteins 94
reciprocal connectivity 69
rejection system 57, 73
renin 33, 34, 87–91, 95, 112, 119, 131
representation of salty taste 9
reproduction 2, 31, 45–7
'reward' neurons 71
rostral brain sites 57
ruminants 103
running speed 22, 23, 37, 39, 75, 132

sagittal section of brain 117
salience of the specialized salt taste 86
saliva 59–60
salt intake 48
salt licks 85
salt pegs 7
'salt pica' 85
salt seeking behavior 6–30, 138
salt taste development 70
'salt'-best fibers 74
saltiness of time (Shakespeare) 15
salty tasting solutes 37
same-place group experiment 18
satiety, salt satiety 78–9
searching for something salty, gustation
    81–6
sensitivity, sodium deprivation 62
sensors, sodium sensors 118
sexual dimorphism 2, 31, 45–7, 54
sheep 11, 22, 23, 35, 37, 40, 42–3, 48, 77–8,
    90, 98, 104, 105, 135
sodium and psychology 1
sodium deprived stimulus spaces 72
sodium in oral cavity 60
sodium reservoir 101–3
sodium-specific transduction mechanism 63
solitary nucleus 66, 91, 107–8, 121, 134, 135
specialist–generalist gustatory fiber
    distinction 86
species differences 77–8
splanchic nerve 134
steroid hormones 293, 93, 94, 125
stimuli 63, 71, 72
stomach 76

stressful conditions 42, 44
stria terminalis 124
subfornical organ 120
surplus of sodium 102–3
sweet and nonsweet stimuli 24, 72, 81–2
'sweet best' neurons 71
synergism 38, 90, 126
synthesia, verse 110

tachykinins 131
tag for minerals 86
taste
    aversion 24
    salt taste development 70
    searching for something salty 81–6
    salt seeking behavior 19, 30
    taste version 76, 132
    taste-elicited consummatory responses 24
    taste reactivity 25
    taste sensitivity 3
    taste transduction 64
    taste experience 28
testosterone 52
'tetrapod novelty' 31
thalamic–cortical sensory pathway 68
thalamo–cortical connection 68
thirst 92, 99
threshold of detection 58–65
thyroid 106
time, animal learning 17
tongue 73
transport of sodium 103
tuning, gustation 69

vagotomy 76, 100–1
vagus nerve 99, 134
venous return to the heart 98
ventral basal thalamus 68
ventral forebrain sites 91
ventral lamina terminalis 118–20
ventral pathway 68–9
ventricular infusion of angiotensin 33
ventromedial nucleus 97
visceral functions 1, 69, 121
vitamins 8, 85, 86, 138

water
    water consumption 100
    water deprivation 106
    water intake 34–5, 128
    water loading 59
West Indian blacks 70
'whole body physiology' 4, 87–109
'wisdom of the body' 4